EVERYDAY ZEN

LOVE AND WORK

Charlott

Edited by Steve Smith

HarperOne
An Imprint of HarperCollinsPublishers

HarperOne

EVERYDAY ZEN: *Love and Work.* Copyright © 1989 by Charlotte Joko Beck. All rights reserved. Printed in the United States of America. No part of this book may be used or reproduced in any manner whatsoever without written permission except in the case of brief quotations embodied in critical articles and reviews. For information address HarperCollins Publishers, 195 Broadway, New York, NY 10007.

Library of Congress Cataloging-in-Publication Data is available.

ISBN 978–0–06–128589–9

23 24 25 26 27 LBC 30 29 28 27 26

Contents

Preface

Successful living means functioning well in love and work, declared Sigmund Freud. Yet most Zen teaching derives from a monastic tradition that is far removed from the ordinary world of romantic and sexual love, family and home life, ordinary jobs and careers. Few Western students of Zen live apart in traditionally structured monastic communities. Most are preoccupied with the same tasks as everyone else: creating or dissolving a relationship, changing diapers, negotiating a mortgage, seeking a job promotion. But the Zen centers that serve such students often retain an aura of esoteric specialness and separateness. Black robes, shaved heads, and traditional monastic rituals may reinforce the impression of Zen as an exotic alternative to ordinary life, rather than ordinary life itself, lived more fully. Because the images and experiences of classical Zen arose out of monasticism, classically trained teachers of Zen are not always able to speak to the actual life issues of their twentieth-century Western students. They may unwittingly encourage an escapist response to those issues, a retreat from problems of real life under the guise of seeking special, overwhelming experiences. If Zen is to become integrated into Western culture, it requires a Western idiom: "Chop wood, carry water" must somehow become, "Make love, drive freeway."

Yet so long as people seek to awaken to themselves and to their life as it is—to the immediacy of this very moment—the spirit of Zen will appear. On a quiet side street of a San Diego suburb, in a small and unprepossessing tract home, just this noiseless burgeoning is under way. Exploring ordinary human relationships, unraveling the dilemmas of career and ambition—bringing Zen to love and work—is the core of "an amazingly pure and lively Zen"* as taught by Charlotte Joko Beck.

---

* Sources for quotations are listed at the end of this book.

Joko Beck is an American Zen original. Born in New Jersey, educated in public schools and at Oberlin Conservatory of Music, Joko (then Charlotte) married and began to raise a family. When the marriage dissolved she supported herself and her four children as a teacher, secretary, and later as an administrative assistant in a large university department. Not until well into her forties did Joko begin the practice of Zen, with Maezumi Roshi (then Sensei) of Los Angeles, and later with Yasutani Roshi and Soen Roshi. For years she commuted regularly from San Diego to the Zen Center of Los Angeles (ZCLA). Her natural aptitude and persistent diligence enabled her to progress steadily; and she found herself increasingly drawn into teaching, as other students recognized her maturity, clarity, and compassion. Joko was eventually designated Maezumi Roshi's third Dharma Heir and, in 1983, she moved to the Zen Center of San Diego, where she now lives and teaches.

As an American woman whose life was well formed before she began to practice, Joko is free from the patriarchal trappings of traditional Japanese Zen. Devoid of pretension or self-importance, she teaches a form of Zen that manifests the ancient Chan principle of *wu shih*—"nothing special." Since moving her practice to San Diego, she no longer shaves her head, and seldom uses robes or her titles. She and her students are evolving an indigenous American Zen that, while still rigorous and disciplined, is adapted to Western temperaments and ways of life.

Joko's dharma talks are models of incisive simplicity and tart common sense. Her own lifetime of personal struggle and growth and her many years of responding with matter-of-fact compassion to her students' traumas and confusions have generated both uncommon psychological insight, and a teacher's gift for apt phrases and telling images. Her teaching is highly pragmatic, less concerned with the concentrative pursuit of special experiences than with the development of insight into the whole of life. Vividly aware that powerful spiritual openings that are artificially induced do not insure an orderly and compassionate life (and may even be harmful), Joko is skeptical of all muscular efforts to overpower one's resistance and find shortcuts to salvation. She favors a slower but healthier, more responsible development of the whole personality, in which psychological barriers are addressed rather than bypassed. One of her students, Elihu Genmyo Smith, reflects her thinking in his description:

There is another way of practice, which I call "working with everything," including emotions, thoughts, sensations, and feelings. Instead of pushing or keeping them away with our mind like an iron wall, or boring through them with our concentration power, we open ourselves up to them. We develop our awareness of what is occurring moment by moment, what thoughts are arising and passing, what emotions we are experiencing, and so on. Instead of a narrow focused concentration, broad awareness is our concentration.

The point is to become more awake to what is ocurring "inside" and "outside." In sitting we sense what is, and we allow it to go on, not attempting to hold it, analyze it, or push it away. The more clearly we see the nature of our sensations, emotions, and thoughts, the more we are able to see through them naturally.

Operating from a perspective of equality, Joko sees herself as a guide rather than a guru, refusing to be put on a pedestal of any kind. Instead, she shares her own life difficulties, thereby creating a humane environment that empowers her students to find their own way.

The selections included here are edited versions of informal talks recorded during intensive meditation retreats or regular Saturday morning programs. In them Joko frequently refers to *zazen*, the traditional Zen meditation, simply as "sitting." *Sangha* is the Buddhist term for the congregation or community of persons who practice together; the *Dharma* may be approximately translated as "Truth," "learning," or "right living." The retreats, called *sesshin* (from the Japanese for "to connect the mind"), last from two to seven days and are conducted in silence, except for essential communication between teacher and student. Beginning in the early morning hours of each day, with eight or more hours of sitting plus meditative work practice, such retreats are both difficult personal challenges and potentially powerful processes of awakening.

Joko has little patience with romanticized spirituality, idealized "sweetness and light" that seeks to bypass reality and the suffering it brings. She is fond of quoting a line from the *Shōyō Rōku*: "From the withered tree, a flower blooms." Through living each moment as it is, the ego gradually drops away, revealing the wonder of everyday life. Joko travels this path with us, and her words, extraordinary in their very ordinariness, help to point the way: elegant wisdom in plain clothes.

Acknowledgments

Numerous students and friends of Joko have eagerly supported the creation of this book, helping to make my own work a labor of love. Generations of anonymous typists who first transcribed Joko's talks into written form deserve much credit; though I cannot begin to thank all of them, they know who they are. In the early stages of planning Rhea Loudon offered important encouragement, and has provided key support throughout. Conversations with Larry Christensen, Anna Christensen, Elihu Genmyo Smith, and Andrew Taido Cooper helped to shape the book; I am grateful for their knowledgeable support. Arnold Kotler of Parallax Press offered generous counsel and astute advice at another critical juncture. Elizabeth Hamilton, who for years has sustained the day-to-day operations of the Zen Center of San Diego with an extraordinary commitment of time, energy, and love, deserves warm thanks for her contributions. My colleague, Christopher Ives, has been an important source of scholarly information and moral support. Professor Masao Abe's thorough knowledge of sources aided me in the final stages of preparation. Pat Padilla has provided rare secretarial service, typing and retyping the manuscript rapidly and impeccably, always with cheerful helpfulness and interest; her contribution is central. Lenore Friedman's fine informal picture of Joko and her teachings, in *Meetings with Remarkable Women: Buddhist Teachers in America* (Boston and London: Shambhala, 1987) was helpful to me in writing the Preface.

The vision of John Loudon at Harper & Row has ultimately made the book possible. With the help of his assistant, Kathryn Sweet, he has guided the book to completion. Their open friendliness and competent support have made my contacts with Harper & Row a delight.

And despite her reservations about the publication of a book that might be expected to attract additional attention to herself and to the small Zen Center where she teaches, Joko has been unfailingly generous and gracious with me – both in my preparation of the manuscript, and in my faltering efforts to walk my own path. To her I owe the greatest debt of all.

Steve Smith
Berkeley, California
February 1988

I. BEGINNINGS

Beginning Zen Practice

My dog doesn't worry about the meaning of life. She may worry if she doesn't get her breakfast, but she doesn't sit around worrying about whether she will get fulfilled or liberated or enlightened. As long as she gets some food and a little affection, her life is fine. But we human beings are not like dogs. We have self-centered minds which get us into plenty of trouble. If we do not come to understand the error in the way we think, our self-awareness, which is our greatest blessing, is also our downfall.

To some degree we all find life difficult, perplexing, and oppressive. Even when it goes well, as it may for a time, we worry that it probably won't keep on that way. Depending on our personal history, we arrive at adulthood with very mixed feelings about this life. If I were to tell you that your life is already perfect, whole, and complete just as it is, you would think I was crazy. Nobody believes his or her life is perfect. And yet there is something within each of us that basically knows we are boundless, limitless. We are caught in the contradiction of finding life a rather perplexing puzzle which causes us a lot of misery, and at the same time being dimly aware of the boundless, limitless nature of life. So we begin looking for an answer to the puzzle.

The first way of looking is to seek a solution outside ourselves. At first this may be on a very ordinary level. There are many people in the world who feel that if only they had a bigger car, a nicer house, better vacations, a more understanding boss, or a more interesting partner, then their life would work. We all go through that one. Slowly we wear out most of our "if onlies." "If only I had this, or that, then my life would work." Not one of us isn't, to some degree, still wearing out our "if onlies." First of all we wear out those on the gross levels. Then we shift our search to more subtle levels. Finally, in looking for the thing outside of ourselves that we hope is going to complete us, we turn to a spiritual discipline. Unfortunately we tend to bring into this new search the same orientation as before. Most people who come to the Zen Center don't think a Cadillac will do it, but they think that enlightenment will. Now they've got a new cookie, a new "if only." "If only I could understand what realization is all about, I would be happy." "If only I could have at least a little enlightenment experience, I would be happy." Coming into a practice like Zen, we bring our usual notions that we are going to get somewhere—become enlightened—and get all the cookies that have eluded us in the past.

Our whole life consists of this little subject looking outside itself for an object. But if you take something that is limited, like body and mind, and look for something outside it, that something becomes an object and must be limited too. So you have something limited looking for something limited and you just end up with more of the same folly that has made you miserable.

We have all spent many years building up a conditioned view of life. There is "me" and there is this "thing" out there that is either hurting me or pleasing me. We tend to run our whole life trying to avoid all that hurts or displeases us, noticing the objects, people, or situations that we think will give us pain or pleasure, avoiding one and pursuing the other. Without exception, we all do this. We remain separate from our life, looking at it, analyzing it, judging it, seeking to answer the questions, "What am I going to get out of it? Is it going to give me pleasure or comfort or should I run away from it?" We do this from morning until night. Underneath our nice, friendly facades there is great unease. If I were to scratch below the surface of anyone I would find fear, pain, and anxiety running amok. We all have ways to cover them up. We overeat, over-drink, overwork; we watch too much television. We are always doing something to cover up our basic existential anxiety. Some people live that way until the day they die. As the years go by, it gets worse and worse. What might not look so bad when you are twenty-five looks awful by the time you are fifty. We all know people who might as well be dead; they have so contracted into their limited viewpoints that it is as painful for those around them as it is for themselves. The flexibility and joy and flow of life are gone. And that rather grim possibility faces all of us, unless we wake up to the fact that we need to work with our life, we need to practice. We have to see through the mirage that there is an "I" separate from "that." Our practice is to close the gap. Only in that instant when we and the object become one can we see what our life is.

Enlightenment is not something you achieve. It is the absence of something. All your life you have been going forward after something, pursuing some goal. Enlightenment is dropping all that. But to talk about it is of little use. The practice has to be done by each individual. There is no substitute. We can read about it until we are a thousand years old and it won't do a thing for us. We all have to practice, and we have to practice with all of our might for the rest of our lives.

What we really want is a natural life. Our lives are so unnatural that to do a practice like Zen is, in the beginning, extremely difficult. But once we begin to get a glimmer that the problem in life is not outside ourselves, we have begun to walk down this path. Once that awakening starts, once we begin to see that life can be more open and joyful than we had ever thought possible, we want to practice.

We enter a discipline like Zen practice so that we can learn to live in a sane way. Zen is almost a thousand years old and the kinks have been worked out of it; while it is not easy, it is not insane. It is down to earth and very practical. It is about our daily life. It is about working better in the office, raising our kids better, and having better relationships. Having a more sane and satisfying life must come out of a sane, balanced practice. What we want to do is to find some way of working with the basic insanity that exists because of our blindness.

It takes courage to sit well. Zen is not a discipline for everyone. We have to be willing to do something that is not easy. If we do it with patience and perseverance, with the guidance of a good teacher, then gradually our life settles down, becomes more balanced. Our emotions are not quite as domineering. As we sit, we find that the primary thing we must work with is our busy, chaotic mind. We are all caught up in frantic thinking and the problem in practice is to begin to bring that thinking into clarity and balance. When the mind becomes clear and balanced and is no longer caught by objects, there can be an opening—and for a second we can realize who we really are.

But sitting is not something that we do for a year or two with the idea of mastering it. Sitting is something we do for a lifetime. There is no end to the opening up that is possible for a human being. Eventually we see that we are the limitless, boundless ground of the universe. Our job for the rest of our life is to open up into that immensity and to express it. Having more and more contact with this reality always brings compassion for others and changes our daily life. We live differently, work differently, relate to people differently. Zen is a lifelong study. It isn't just sitting on a cushion for thirty or forty minutes a day. Our whole life becomes practice, twenty-four hours a day.

Now I would like to answer some questions about Zen practice and its relation to your life.

STUDENT: Would you expand upon the idea of letting go of thoughts that occur during meditation?

JOKO: I don't think that we ever let go of anything. I think what we do is just wear things out. If we start forcing our mind to do something, we are right back into the dualism that we are trying to get out of. The best way to let go is to notice the thoughts as they come up and to acknowledge them. "Oh, yes, I'm doing that one again"—and without judging, return to the clear experience of the present moment. Just be patient. We might have to do it ten thousand times, but the value for our practice is the constant return of the mind into the present, over and over and over. Don't look for some wonderful place where thoughts won't occur. Since the thoughts basically are not real, at some point they get dimmer and less imperative and we will find there are periods when they tend to fade out because we see they are not real. They will just wither away in time without our quite knowing how it happened. Those thoughts are our attempt to protect ourselves. None of us really wants to give them up; they are what we are attached to. The way we can eventually see their unreality is by just letting the movie run. After we have seen the same movie five hundred times it gets boring, frankly!

There are two kinds of thoughts. There is nothing wrong with thinking in the sense of what I call "technical thinking." We have to think in order to walk from here to the corner or to bake a cake or to solve a physics problem. That use of the mind is fine. It isn't real or unreal; it is just what it is. But opinions, judgments, memories, dreaming about the future—ninety percent of the thoughts spinning around in our heads have no essential reality. And we go from birth to death, unless we wake up, wasting most of our life with them. The gruesome part of sitting (and it is gruesome, believe me) is to begin to see what is really going on in our mind. It is a shocker for all of us. We see that we are violent, prejudiced, and selfish. We are all those things because a conditioned life based on false thinking leads to these states. Human beings are basically good, kind, and compassionate, but it takes hard digging to uncover that buried jewel.

STUDENT: You said that as time goes on the ups and downs, the upsets, begin to dwindle until they just peter out?

JOKO: I am not implying that there will not be upsets. What I mean is that when we get upset, we don't hold onto it. If we become angry, we are just angry for a second. Others may not even be aware of it. That is all there is to it. There is no clinging to the anger, no mental spinning with it. I don't mean that years of practice leave us like a zombie. Quite the opposite. We really have more genuine emotions, more feeling for people. We are not so caught up in our own inner states.

STUDENT: Would you please comment on our daily work as part of our practice?

JOKO: Work is the best part of Zen practice and training. No matter what the work is, it should be done with effort and total attention to what's in front of our nose. If we are cleaning the oven, we should just totally do that and also be aware of any thoughts that interrupt the work. "I hate to clean the oven. Ammonia smells! Who likes to clean the oven, anyway? With all my education I shouldn't have to do this." All those are extra thoughts that have nothing to do with cleaning the oven. If the mind drifts in any way, return it to the work. There is the actual task we are doing and then there are all the considerations we have about it. Work is just taking care of what needs to be done right now, but very few of us work that way. When we practice patiently, eventually work begins to flow. We just do whatever needs to be done.

No matter what your life is, I encourage you to make it your practice.

Practicing This Very Moment

I'd like to talk about the basic problem of sitting. Whether you've been sitting a short time or for ten years, the problem is always the same.

When I went to my first sesshin many years ago, I couldn't decide who was crazier—me or the people sitting around me. It was terrible! The temperature was almost 105 degrees every day of the week, I was covered with flies, and it was a noisy, bellowing sesshin. I was completely upset and baffled by the whole thing. But once in a while I'd go in to see Yasutani Roshi, and there I saw something that kept me sitting. Unfortunately, the first six months or year of sitting are the hard ones. You have to face confusion, doubts, problems; and you haven't been sitting long enough to feel the real rewards.

But the difficulty is natural, even good. As your mind slowly goes through all of these things, as you sit here, confusing and ridiculous as it may seem, you're learning a tremendous amount about yourself. And this can only be of value to you. Please continue to sit with a group as often as you can, and see a good teacher as often as you can. If you do that, in time, this practice will be the best thing in your life.

It doesn't matter what our practice is called: following the breath, shikan-taza, koan study; basically, we're all working on the same issues: "Who are we? What is our life? Where did we come from? Where do we go?" It's essential to living a whole human life that we have some insight. So first I'd like to talk about the basic task of sitting—and, in talking about it, realize that talking is not it. Talking is just the finger pointing at the moon.

In sitting we are uncovering Reality, Buddha-nature, God, True Nature. Some call it "Big Mind." Words for it that are particularly apt for the way I want to approach the problem tonight are "this very moment."

The Diamond Sutra says, "The past is ungraspable, the present is ungraspable, the future is ungraspable." So all of us in this room; where are we? Are we in the past? No. Are we in the future? No. Are we in the present? No, we can't even say we're in the present. There's nothing we can point to and say, "This is the present," no boundary lines that define the present. All we can say is, "We are this very moment." And because there's no way of measuring it, defining it, pinning it down, even seeing what it is, it's immeasurable, boundless, infinite. It's what we are.

Now, if it's as simple as that, what are we all doing here? I can say, "This very moment." That sounds easy doesn't it? Actually it's not. To really see it is not so easy, or we wouldn't all be doing this.

Why isn't it easy? Why can't we see it? And what is necessary so that we can see it? Let me tell you a little story.

Many years ago I was a piano major at Oberlin Conservatory. I was a very good student; not outstanding, but very good. And I very much wanted to study with one teacher who was undoubtedly the best. He'd take ordinary students and turn them into fabulous pianists. Finally I got my chance to study with *the* teacher.

When I went in for my lesson I found that he taught with two pianos. He didn't even say hello. He just sat down at his piano and played five notes, and then he said, "You do it." I was supposed to play it just the way he played it. I played it—and he said, "No." He played it again, and I played it again. Again he said, "No." Well, we had an hour of that. And each time he said, "No."

In the next three months I played about three measures, perhaps half a minute of music. Now I had thought I was pretty good: I'd played soloist with little symphony orchestras. Yet we did this for three months, and I cried most of those three months. He had all the marks of a real teacher, that tremendous drive and determination to make the student see. That's why he was so good. And at the end of three months, one day, he said, "Good." What had happened? Finally, I had learned to listen. And as he said, if you can hear it, you can play it.

What had happened in those three months? I had the same set of ears I started with; nothing had happened to my ears. What I was playing was not technically difficult. What had happened was that I had learned to listen for the first time . . . and I'd been playing the piano for many years. I learned to pay attention. That was why he was such a great teacher: he taught his students to pay attention. After working with him they really heard, they really listened. When you can hear it, you can play it. And finished, beautiful pianists would finally come out of his studio.

It's that kind of attention which is necessary for our Zen practice. We call it samadhi, this total oneness with the object. But in my story that attention was relatively easy. It was with an object that I liked. This is the oneness of any great art, the great athlete, the person who passes well on the football field, the person who does well on the basketball court, anybody like that who has to learn to pay attention. It's that kind of samadhi.

Now that's one kind, and it's valuable. But what we have to do in Zen practice is much harder. We have to pay attention to this very moment, the totality of what is happening right now. And the reason we don't want to pay attention is because it's not always pleasant. It doesn't suit us.

As human beings we have a mind that can think. We remember what has been painful. We constantly dream about the future, about the nice things we're going to have, or are going to happen to us. So we filter anything happening in the present through all that: "I don't like that. I don't have to listen to that. And I can even forget about it and start dreaming of what's going to happen." This goes on constantly: spinning, spinning, spinning, always trying to create life in a way that will be pleasant, that would make us safe and secure, so we feel good.

But when we do that we never see this right-here-now, this very moment. We can't see it because we're filtering. What's coming in is something quite different. Just ask any ten people who read this book. You'll find they all tell you something different. They'll forget the parts that don't quite catch them, they'll pick up something else, and they'll even block out the parts they don't like. Even when we go to our Zen teacher we hear only what we want to hear. Being open to a teacher means not just hearing what you want to hear, but hearing the whole thing. And the teacher's not there simply to be nice to you.

So the crux of zazen is this: all we must do is constantly to create a little shift from the spinning world we've got in our heads to right-here-now. That's our practice. The intensity and ability to be right-here-now is what we have to develop. We have to be able to develop the ability to say, "No, I won't spin off up here" to make that choice. Moment by moment our practice is like a choice, a fork in the road: we can go this way, we can go that way. It's always a choice, moment by moment, between our nice world that we want to set up in our heads and what really is. And what really is, at a Zen sesshin, is often fatigue, boredom, and pain in our legs. What we learn from having to sit quietly with that discomfort is so valuable that if it didn't exist, it should. When you're in pain, you can't spin off. You have to stay with it. There's no place to go. So pain is really valuable.

Our Zen training is designed to enable us to live comfortable lives. But the only people who live comfortably are those who learn not to dream their lives away, but to be with what's right-here-now, no matter what it is: good, bad, nice, not nice, headache, being ill, being happy. It doesn't make any difference.

One mark of a mature Zen student is a sense of groundedness. When you meet one you sense it. They're with life as it's really happening, not as a fantasy version of it. And of course, the storms of life eventually hit them more lightly. If we can accept things just the way they are, we're not going to be greatly upset by anything. And if we do become upset it's over more quickly.

Let's look at the sitting process itself. What we need to do is to be with what's happening right now. You don't have to believe me; you can experiment for yourself. When I am drifting away from the present, what I do is listen to the traffic. I make sure there's nothing I miss. Nothing. I just really listen. And that's just as good as a koan, because it's what's happening this very moment. So as Zen students you have a job to do, a very important job: to bring your life out of dreamland and into the real and immense reality that it is.

The job is not easy. It takes courage. Only people who have tremendous guts can do this practice for more than a short time. But we don't do it just for ourselves. Perhaps we do at first; that's fine. But as our life gets grounded, gets real, gets basic, other people immediately sense it, and what we are begins to influence everything around us.

We are, actually, the whole universe. But until you see that clearly, you have to work with what your teacher tells you to work with, having some faith in the total process. It's not only faith, it's also something like science. Others before you have done the experiment, and they've had some results from that. About all you can do is say, "Well, at least I can try the experiment. I can do it. I can work hard." That much any of us can do.

The Buddha is nothing but exactly what you are, right now: hearing the cars, feeling the pain in your legs, hearing my voice; that's the Buddha. You can't catch hold of it; the minute you try to catch it, it's changed. Being what we are at each moment means, for example, fully being our anger when we are angry. That kind of anger never hurts anybody because it's total, complete. We really feel this anger, this knot in our stomach, and we're not going to hurt anybody with it. The kind of anger that hurts people is when we smile sweetly and underneath we're seething.

When you sit, don't expect to be noble. When we give up this spinning mind, even for a few minutes, and just sit with what is, then this presence that we are is like a mirror. We see everything. We see what we are: our efforts to look good, to be first, or to be last. We see our anger, our anxiety, our pomposity, our so-called spirituality. Real spirituality is just being with all that. If we can really be with Buddha, who we are, then it transforms.

Shibayama Roshi said once in sesshin, "This Buddha that you all want to see, this Buddha is very shy. It's hard to get him to come out and show himself." Why is that? Because the Buddha is ourselves, and we'll never see the Buddha until we're no longer attached to all this extra stuff. We've got to be willing to go into ourselves honestly. When we can be totally honest with what's happening right now, then we'll see it. We can't have just a piece of the Buddha. Buddhas come whole. Our practice has nothing to do with, "Oh, I should be good, I should be nice, I should be this . . . or that." I *am* who I am right now. And that very state of being is the Buddha.

I once said something in the zendo that upset a lot of people: I said, "To do this practice, we have to give up hope." Not many were happy about that. But what did I mean? I mean that we have to give up this idea in our heads that somehow, if we could only figure it out, there's some way to have this perfect life that is just right for us. Life is the way it is. And only when we begin to give up those maneuvers does life begin to be more satisfactory.

When I say to give up hope, I don't mean to give up effort. As Zen students we have to work unbelievably hard. But when I say hard, I don't mean straining and effort; it isn't that. What is hard is this choice that we repeatedly have to make. And if you practice hard, come to a lot of sesshins, work hard with a teacher, if you're willing to make that choice consistently over a period of time, then one day you'll get your first little glimpse. This first little glimpse of what this very moment is. And it might take one year, two years, or ten years.

Now that's the beginning. That one little glimpse takes a tenth of a second. But just that isn't enough. The enlightened life is seeing that all the time. It takes years and years and years of work to transform ourselves to the point where we can do that.

I don't mean to sound discouraging. You might feel you probably don't have enough years left to do it. But that's not the point. At every point in our practice it's perfect. And as we practice life steadily becomes more fulfilling, more satisfactory, better for us, better for other people. But it's a long, long continuum. People have some silly idea that they're going to be enlightened in two weeks.

Already we are the Buddha. There's just no doubt about that. How could we be anything else? We're all right here now. Where else could we be? But the point is to realize clearly what that means; this total oneness; this harmony; and to be able to express that in our lives. That's what takes endless work and training. It takes guts. It's not easy. It takes a real devotion to ourselves and to other people.

Now of course, as we practice, all these things grow, even the guts. We have to sit with pain and we hate it. I don't like it either. But as we patiently just sit our way through that, something builds within us. Working with a good teacher, seeing what she or he is, we are slowly transformed in this practice. It's not by anything we think, not by something we figure out in our heads. We're transformed by what we do. And what is it that we do? We constantly make that choice. We give up our ego-centered dreams for this reality that we really are.

We may not understand it at first; it may be confusing. When I first heard talks by teachers I'd think, "What *are* they talking about?" But have enough faith to just do your practice: Sit every day. Go through the confusion. Be very patient. And respect yourself for doing this practice. It's not easy. Anyone who sits through a Zen sesshin is to be congratulated. I'm not trying to be hard on you; I think people who come to this practice are amazing people. But it's your job to take that quality you have and work with it.

We're all just babies. The extent to which we can grow is boundless. And eventually, if we're patient enough, and work hard enough, we have some possibility of making a real contribution to the world. In this oneness that we finally learn to live in, that's where the love is; not some kind of a soupy version, but a love with real strength. We want it for our lives, and we want it for other people's lives. We want it for our children, our parents, our friends. So it's up to us to do the work.

So that's the process. Whether we choose to do it is up to us. The process may not be clear to many of you; it takes years before it becomes clear, so that you really know what you're doing. Just do the best you can. Stay with your sitting. Come to sesshin, come to sit, and let's all do our best. It's really important: this total transformation of the quality of human life is the most important thing we can do.

Authority

After years of talking to many, many people I'm still amazed that we make such a problem of our life and practice. And there is *no* problem. But saying that is one thing, seeing it is quite another. The last words of the Buddha were, "Be a lamp unto yourself." He didn't say, "Go running to this teacher or that teacher, to this center or that center"—he said, "Look—be a lamp unto yourself."

What I want to discuss here is the problem of "authority." Usually we're either an authority to others (telling them what to do), or we're seeking someone to be an authority for us (telling us what to do). And yet we would never be looking for an authority if we had any confidence in ourselves and our understanding. Particularly when there is something in our life that is unpleasant or baffling or upsetting, we think we need to go to a teacher or authority who can tell us what to do. I'm always amused that when a new teacher comes to town, everyone goes running to see him or her. I'll tell you how far I'd walk to see a new teacher:

maybe across the room, no farther! It isn't because I have no interest in this person; it's just that there is *no one* who can tell me about my life except—who? There *is* no authority outside of my experience.

But you may say, "Well, I need a teacher who can free me from my suffering. I'm hurting and I don't understand it. I need someone who can tell me what to do, don't I?" No! You may need a guide, you may need it made clear how to practice with your life—what is needed is a guide who will make it clear to you that the authority in your life, your true teacher, is you—and we practice to realize this "you."

There is only one teacher. What is that teacher? Life itself. And of course each one of us is a manifestation of life; we couldn't be anything else. Now life happens to be both a severe and an endlessly kind teacher. It's the only authority that you need to trust. And this teacher, this authority, is everywhere. You don't have to go to some special place to find this incomparable teacher, you don't have to have some especially quiet or ideal situation: in fact, the messier it is, the better. The average office is a great place. The average home is perfect. Such places are pretty messy most of the time—we all know from firsthand experience! That is where the authority, the teacher is.

This is a very radical teaching, not for everyone. People often turn away from such a teaching; they don't want to hear it. What do they want to hear? What do *you* want to hear? Until we're ready (which usually means, until we have suffered and have been willing to learn from the suffering) we're like baby birds in a nest. What do baby birds do? They open their mouths upward and wait to be fed. And we say, "Please stuff your wonderful teaching into me. I'll hold my mouth open, but you put it in." What we are saying is, "When will Mommy and Daddy come? When will a great teacher, a supreme authority, come and stuff me with that which will end my pain, my suffering?" The news is, Mommy and Daddy have already come! Where *are* Mommy and Daddy? Right here. Our life is always here! But since my life may look to me like discomfort, even dreariness, loneliness, depression, if I actually were to face that (life as it is), who would want that? Almost no one. But when I can begin to experience this very moment, the true teacher—when I can honestly *be* each moment of my life, what I think, feel—this experiencing will settle itself into "just this," the joyful samadhi of life, the word of God. And that *is* Zen practice, and we don't even have to use the word "Zen."

This Mommy and Daddy that we've been waiting for are already here—right here. We can't avoid the authority even if we want to. When we go to work, it's right there; when we're with our friends, it's right there; when with our family, it's right there. "Do zazen constantly; pray constantly." If we understand that each moment of our life is the teacher, we can't avoid doing that. If we *truly* are each moment of our life there is no room for an outside influence or authority. Where could it be? When I am just my own suffering where is the authority? The attention, the experiencing *is* the authority, and it is also the clarification of the action to be done.

There is one final little illusion that we all tend to play with in this question of authority, and it is, "Well, I'll be my *own* authority, thank you. No one is going to tell *me* what to do." What is the falsity in this? "I'll be my own authority! I'll develop my own concepts about life, my own ideas of what Zen practice is"—we're all full of this nonsense. If I attempt to be my own authority (in this narrow sense), I am just as much a slave as if I let someone else be the authority. But if you are not the authority and I am not the authority, then what? We've already talked about this but, if it's not understood clearly, we may be floundering in quicksand. How do you see it?

The Bottleneck of Fear

The limitations of life are present at conception. In the genetic factors themselves are limitations: we are male or female, we have tendencies to certain diseases or bodily weaknesses. All the genetic strands come together to produce a certain temperament. It's evident to any mother carrying a child that there is a tremendous difference between babies even before birth. But for our purposes we may begin with the baby at birth. To adult eyes a newborn baby seems open and unconditioned. In its early weeks of life, a baby's imperative is to survive. Just listen to a newborn baby scream—it can easily run the whole household. I can't think of anything else that has the riveting quality of the screaming of a newborn. When I hear that sound I want to do something, anything, to stop it. But it doesn't take long for baby to learn that, in spite of its strenuous efforts, life isn't always pleasant. I remember dropping my oldest son on his head when he was six weeks old. I thought I was such an expert new mother, but he was soapy and. . . .

Very early we all begin our attempt to protect ourselves against the threatening occurrences that pop up regularly. In the fear caused by them, we begin to contract. And the open, spacious character of our young life feels pushed through a funnel into a bottleneck of fear. Once we begin to use language the rapidity of this contracting increases. And particularly as our intelligence grows, the process becomes really speedy: now we not only try to handle the threat by storing it in every cell of our body, but (using memory) we relate each new threat to all of the previous ones— and so the process compounds itself.

We are all familiar with the process of conditioning: suppose that when I was a little girl a redheaded, big and tough, five-year-old boy snatched my favorite toy; I was frightened—and conditioned. Now every time a person with red hair passes through my life I feel uneasy for no obvious reason. Could we say then that conditioning is the problem? No, not exactly. Conditioning, even when often repeated, fades in time. For that reason, the person who says, "If you knew what my life has been like, it's no wonder I'm such a mess—I'm so conditioned by fear, it's hopeless," isn't grasping the real problem. What *is* true is that—yes—we all have been repeatedly conditioned and, under the influence of those incidents, we slowly revise our ideas of who we are. Having been threatened in our openness and spaciousness, we make a decision that our self is the contraction of fear. I revise my notions of myself and the world and define a new picture of myself; and whether that new picture is one of compliance or noncompliance or withdrawal doesn't make much difference. What makes the difference is my blind decision that I now must fulfill my contracted picture of myself in order to survive.

The bottleneck of fear isn't caused by the conditioning, but by the decision about myself I have reached based on that conditioning. Fortunately, because that decision is composed of our thoughts and reflected in bodily contraction, it can be my teacher when I experience myself in this present moment. I don't necessarily need an intellectual knowledge of what my conditioning has been, although this can be helpful. What I do need to know is what sorts of thoughts I persist in entertaining right now, today, and what bodily contractions I have right now, today. In noticing the thoughts and in experiencing the bodily contractions (doing zazen), the bottleneck of fear is illuminated. And as I do this my false identification with a limited self (the decision) slowly fades. More and more I can be who I truly am: a no-self, an open and spacious response to life. My true self, so long deserted and forgotten, can function, now that I can see that the bottleneck of fear is an illusion.

I'm reminded at this point of the two famous verses about a mirror (one by a monk who was a fine student of the Fifth Patriarch, and the other by an unknown who would become the Sixth Patriarch). These verses were composed so that the Fifth Patriarch could judge whether or not the writer had true realization. The monk's verse (the one that was not accepted by the Fifth Patriarch as the truth) stated that practice consists of polishing the mirror; in other words, by removing the dust of our deluded thoughts and actions from the mirror, it can shine (we are purified). The other verse (the one that revealed to the Fifth Patriarch the deep understanding of the man he would choose as his successor) stated that from the very beginning "there is no mirror-stand, no mirror to polish, and no place where dust can cling. . ."

Now while the verse of the Sixth Patriarch is the true understanding, the paradox for us is that we have to practice with the verse that was *not* accepted: we do have to polish the mirror; we do have to be aware of our thoughts and actions; we do have to be aware of our false reactions to life. Only by doing so can we see that from the beginning the bottleneck of fear *is* an illusion. And it is obvious that we do not have to struggle to rid ourselves of an illusion. But we can't and won't know that unless we relentlessly polish the mirror.

Sometimes people say, "Well, there's nothing that need be done. No practice (polishing) is necessary. If you see clearly enough, such practice is nonsense." Ah. . . but we *don't* see clearly enough and, when we fail to see clearly, we create merry mayhem for ourselves and others. We do have to practice, we do have to polish the mirror, until we know in our guts the truth of our life. Then we can see that from the very beginning, nothing was needed. Our life is always open and spacious and fruitful. But let's not fool ourselves about the amount of sincere practice we must do before we see this as clearly as the nose on our face.

What I am presenting to you is really an optimistic view of practice, even though at times the doing will be discouraging and difficult. But again the question is, do we have a lot of choice? Either we die—because if we remain very long in the bottleneck of fear we will be strangled to death—or we slowly gain comprehension by experiencing the bottleneck and going through it. I don't think we have a lot of choice. How about you?

II. PRACTICE

What Practice Is Not

Many people practice and have strong ideas of what practice *is*. What I want to do is to state (from my point of view) what practice is *not*.

First, practice is not about producing psychological change. If we practice with intelligence, psychological change *will* be produced; I'm not questioning that—in fact, it's wonderful. I am saying that practice is not done in order to produce such change.

Practice is not about intellectually knowing the physical nature of reality, what the universe consists of, or how it works. And again, in serious practice, we will tend to have some knowledge of such matters. But that is not what practice is.

Practice is not about achieving some blissful state. It's not about having visions. It's not about seeing white lights (or pink or blue ones). All of these things may occur, and if we sit long enough they probably will. But that is not what practice is about.

Practice is not about having or cultivating special powers. There are many of these and we all have some of them naturally; some people have them in extra measure. At the Zen Center of Los Angeles (ZCLA) I sometimes had the useful ability to see what was being served for dinner two doors away. If they were having something I didn't like, I didn't go. Such abilities are little oddities, and again they are not what practice is about.

Practice is not about personal power or *jōriki*, the strength that is developed in years of sitting. Again, *jōriki* is a natural byproduct of zazen. And again it is not the way.

Practice is not about having nice feelings, happy feelings. It's not about feeling good as opposed to feeling bad. It's not an attempt to be anything special or to feel anything special. The product of practice or the point of practice or what practice is about is not to be always calm and collected. Again, we tend to be much more so after years of practice, but it is not the point.

Practice is not about some bodily state in which we are never ill, never hurt, one in which we have no bothersome ailments. Sitting tends to have health benefits for many people, though in the course of practice there may be months or even years of health disasters. But again, seeking perfect health is not the way; although by and large, over time, there will be a beneficial effect on health for most people. But no guarantees!

Practice is not about achieving an omniscient state in which a person knows all about everything, a state in which a person is an authority on any and all worldly problems. There may be a little more clarity on such matters, but clever people have been known to say and do foolish things. Again, omniscience is not the point.

Practice is not about being "spiritual," at least not as this word is often understood. Practice is not about being *anything*. So unless we see that we cannot aim at being "spiritual," it can be a seductive and harmful objective.

Practice is not about hightlighting all sorts of "good" qualities and getting rid of the so-called "bad" ones. No one is "good" or "bad." The struggle to be good is not what practice is. That type of training is a subtle form of athleticism.

We could continue our listing almost endlessly. Actually anyone in practice has some of these delusions operating. We all hope to change, to get somewhere! That in itself is the basic fallacy. But just contemplating this desire begins to clarify it, and the practice basis of our life alters as we do so. We begin to comprehend that our frantic desire to get better, to "get somewhere," is illusion itself, and the source of suffering.

If our boat full of hope, illusions, and ambition (to get somewhere, to be spiritual, to be perfect, to be enlightened) is capsized, what is that empty boat? Who are we? What, in terms of our lives, can we realize? And what *is* practice?

What Practice Is

Practice is very simple. That doesn't mean it won't turn our life around, however. I want to review what we do when we sit, or do zazen. And if you think you're beyond this, well, you can think you're beyond this.

Sitting is essentially a simplified space. Our daily life is in constant movement: lots of things going on, lots of people talking, lots of events taking place. In the middle of that, it's very difficult to sense what we are in our life. When we simplify the situation, when we take away the externals and remove ourselves from the ringing phone, the television, the people who visit us, the dog who needs a walk, we get a chance—which is absolutely the most valuable thing there is—to face ourselves. Meditation is not about some state, but about the meditator. It's not about some activity, or about fixing something, or accomplishing something. It's about ourselves. If we don't simplify the situation the chance of taking a good look at ourselves is very small—because what we tend to look at isn't ourselves, but everything else. If something goes wrong, what do we look at? We look at what's going wrong, and usually at others we think have made it go wrong. We're looking *out there* all the time, and not at ourselves.

When I say meditation is about the meditator, I do not mean that we engage in self-analysis. That's not it either. So what *do* we do?

Once we have assumed our best posture (which should be balanced, easy), we just sit there, we do zazen. What do I mean by "just sit there"? It's the most demanding of all activities. Usually in meditation we don't shut our eyes. But right now I'd like you to shut your eyes and just *sit* there. What's going on? All sorts of things. A tiny twitch in your left shoulder; a pressure in your side . . . Notice your face for a moment. Feel it. Is it tense anywhere? Around the mouth, around the forehead? Now move down a bit. Notice your neck, just feel it. Then your shoulders, your back, chest, abdominal area, your arms, thighs. Keep feeling whatever you find. And feel your breath as it comes and goes. Don't try to control it, just feel it. Our first instinct is to try to control the breath. Just let your breath be as it is. It may be high in your chest, it may be in the middle, it may be low. It may feel tense. Just experience it as it is. Now just feel all of that. If a car goes by outside, hear it. If a plane flies over, notice that. You might hear a refrigerator going on and off. Just be that. That's all you have to do, absolutely all you have to do: experience that, and just stay with it. Now you can open your eyes.

If you can just do that for three minutes, that's miraculous. Usually after about a minute we begin to think. Our interest in just being with reality (which is what we have just done) is very low. "You mean that is all there is to zazen?" We don't like that. "We're seeking enlightenment, aren't we?" Our interest in reality is extremely low. No, we want to think. We want to worry through all of our preoccupations. We want to figure life out. And so before we know it we've forgotten all about this moment, and we've drifted off into thinking about something: our boyfriend, our girlfriend, our child, our boss, our current fear . . . off we go! There's nothing sinful about such fantasizing except that when we're lost in that, we've lost something else. When we're lost in thought, when we're dreaming, what have we lost? We've lost reality. Our life has escaped us.

This is what human beings do. And we don't just do it sometimes, we do it most of the time. Why do we do that? You know the answer, of course. We do it because we are trying to protect ourselves. We're trying to rid ourselves of our current difficulty, or at least understand it. There's nothing wrong with our self-centered thoughts except that when we identify with them, our view of reality is blocked. So what should we do when the thoughts come up? We should label the thoughts. Be *specific* in your labeling: not just "thinking, thinking" or "worrying, worrying," but a specific label. For example: "Having a thought she's very bossy." "Having a thought that he's very unfair to me." "Having a thought that I never do anything right." Be specific. And if the thoughts are tumbling out so fast that you can't find anything except confusion, then just label the foggy mess "confusion." But if you persist in trying to find a separate thought, sooner or later you will.

When we practice like this, we get acquainted with ourselves, how our lives work, what we are doing with them. If we find that certain thoughts come up hundreds of times, we know something about ourselves that we didn't know before. Perhaps we incessantly think about the past, or the future. Some people always think about events, some people always think about other people. Some people always think about themselves. Some people's thoughts are almost entirely judgments about other people. Until we have labeled for four or five years, we don't know ourselves very well. When we label thoughts precisely and carefully, what happens to them? They begin to quiet down. We don't have to force ourselves to get rid of them. When they quiet down, we return to the experience of the body and the breath, over and over and over. I can't emphasize enough that we don't just do this three times, we do it ten thousand times; and as we do it, our life transforms. That's a theoretical description of sitting. It's very simple; there's nothing complicated about it.

Now let's take a daily life situation. Suppose you work in an aircraft plant, and you're told that the government contract is coming to an end and probably will not be renewed. You tell yourself, "I'm going to lose my job. I'm going to lose my income, I have a family to support. This is terrible!" What happens then? Your mind starts going over and over and over your problem. "What's going to happen? What shall I do?" Your mind spins faster and faster with worry.

Now there's nothing wrong with planning ahead; we have to plan. But when we become upset, we don't just plan; we obsess. We twist the problem around in a thousand ways. If we don't know what it means to practice with our worried thoughts, what happens next? The thoughts produce an emotion and we become even more agitated. All emotional agitation is caused by the mind. And if we let this happen over a period of time, we often become physically sick or mentally depressed. If the mind will not take care of a situation with awareness, the body will. It will help us out. It's as if the body says, "If you won't take care of it, I guess I've got to." So we produce our next cold, our next rash, our next ulcer, whatever is our style. A mind that is not aware will produce illness. That's not a criticism, however. I don't know of anyone who doesn't get ill, including myself. When the desire to worry is strong, we create difficulties. With regular practice, we just do it less. Anything of which we're unaware will have its fruits in our life, one way or another.

From the human point of view, the things that go wrong in our lives are of two kinds. One kind are events outside of ourselves, and the other are things within us, such as physical illness. Both are our practice, and we handle them in the same way. We label all the thoughts that occur around them, and we experience them in our body. The process is sitting itself.

To talk about this sounds really easy. But to do it is horrendously difficult. I don't know anyone who can do it all of the time. I know of some people who can do it much of the time. But when we practice in this way, becoming aware of everything that enters our life (whether internal or external), our life begins to transform. And we gain strength and insight and even live at times in the enlightened state, which simply means experiencing life as it is. It's not a mystery.

If you are new to practice it's important to realize that simply to sit on that cushion for fifteen minutes is a victory. Just to sit with that much composure, just to be there, is fine.

If we were afraid of being in water and didn't know how to swim, the first victory would be just to lower ourselves into the water. The next step might be getting our face wet. If we were expert swimmers the challenge might be whether we can enter our hand into the water at a certain angle as we execute our stroke. Does that mean that one swimmer is better and the other worse? No. Both of them are perfect for where they are. Practice at any stage is just being who we are at that moment. It's not a question of being good or bad, or better or worse. Sometimes after my talks people will say, "I don't understand that." And that's perfect too. Our understanding grows over the years, but at any point we are perfect in being what we are.

We begin to learn that there is only one thing in life we can rely on. What is the one thing in life we can rely on? We might say, "I rely on my mate." We may love our husbands and wives; but we can't ever completely rely on them, because another person (like ourselves) is always to some extent unreliable. There is no person on earth whom we can completely rely on, though we can certainly love others and enjoy them. What then can we rely on? If it's not a person, what is it? What can we rely on in life? I asked somebody once and she said, "Myself." Can you rely on yourself? Self-reliance is nice, but is inevitably limited.

There *is* one thing in life that you can always rely on: life being as it is. Let's talk more concretely. Suppose there is something I want very much: perhaps I want to marry a certain person, or get an advanced degree, or have my child be healthy and happy. But life as it is might be exactly the opposite of what I want. We don't know that we'll marry that certain person. If we do, he might die tomorrow. We may or may not get our advanced degree. Probably we will, but we can't count on that. We can't count on anything. Life is always going to be the way it is. So why can't we rely on that fact? What is so hard about that? Why are we always uneasy? Suppose your living space has just been demolished by an earthquake, and you are about to lose an arm and all your life's savings. Can you then rely on life just as it is? Can you be that?

Trust in things being as they are is the secret of life. But we don't want to hear that. I can absolutely trust that in the next year my life is going to be changed, different, yet always just the way it is. If tomorrow I have a heart attack, I can rely on that, because if I have it, I have it. I can rest in life as it is.

When we make a personal investment in our thoughts we create the "I" (as Krishnamurti would say), and then our life begins not to work. That's why we label thoughts, to take the investment out again. When we've been sitting long enough we can see our thoughts as just pure sensory input. And we can see ourselves moving through the stages preliminary to that: at first we feel our thoughts are real, and out of that we create the self-centered emotions, and out of that we create the barrier to seeing life as it is;

because if we are caught in self-centered emotions we can't see people or situations clearly. A thought in itself is just pure sensory input, an energy fragment. But we fear to see thoughts as they are.

When we label a thought we step back from it, we remove our identification. There's a world of difference between saying, "She's impossible" and "Having a thought that she's impossible." If we persistently label any thought the emotional overlay begins to drop out and we are left with an impersonal energy fragment to which we need not attach. But if we think our thoughts are real we act out of them. And if we act from such thoughts our life is muddled. Again, practice is to work with this until we know it in our bones. Practice is not about achieving a realization in our heads. It has to be our flesh, our bones, ourself. Of course, we have to have life-centered thoughts: how to follow a recipe, how to put on a roof, how to plan our vacation. But we don't need the emotionally self-centered activity that we call thinking. It really isn't thinking, it's an aberration of thinking.

Zen is about an active life, an involved life. When we know our minds well and the emotions that our thinking creates, we tend to see better what our lives are about and what needs to be done, which is generally just the next task under our nose. Zen is about a life of action, not a life of passively doing nothing. But our actions must be based on reality. When our actions are based on our false thought systems (which are based on our conditioning), they are poorly based. When we have seen through the thought systems we can see what needs to be done.

What we are doing is not reprogramming ourselves, but freeing ourselves from all programs, by seeing that they are empty of reality. Reprogramming is just jumping from one pot into another. We may have what we think of as a better programming; but the point of sitting is not to be run by *any* program. Suppose we have a program called "I lack self-confidence." Suppose we decide to reprogram that to "I have self-confidence." Neither of them will stand up very well under the pressures of life, because they involved an "I." And this "I" is a very fragile creation—unreal, actually—and is easily befuddled. In fact there never was an "I." The point is to see that it is empty, an illusion, which is different from dissolving it. When I say that it's empty, I mean that it has no basic reality; it's just a creation of the self-centered thoughts.

Doing Zen practice is never as simple as talking about it. Even students who have a fair understanding of what they're doing at times tend to desert basic practice. Still, when we sit well, everything else takes care of itself. So whether we have been sitting five years or twenty years or are just beginning, it is important to sit with great, meticulous care.

The Fire of Attention

Back in the 1920s, when I was maybe eight or ten years old, and living in New Jersey where the winters are cold, we had a furnace in our house that burned coal. It was a big event on the block when the coal truck rolled up and all this stuff poured down the coal shute into the coal bin. I learned that there were two kinds of coal that showed up in the coal bin: one was called anthracite or hard coal, and the other was lignite, soft coal. My father told me about the difference in the way those two kinds of coal burned. Anthracite burns cleanly, leaving little ash. Lignite leaves lots of ash. When we burned lignite, the cellar became covered with soot and some of it got upstairs into the living room. Mother had something to say about that, I remember. At night my father would bank the fire, and I learned to do this too. Banking the fire means covering it with a thin layer of coal, and then shutting down the oxygen vent to the furnace, so that the fire stays in a slow-burning state. Overnight the house becomes cold, and so in the morning the fire must be stirred up and the oxygen vent opened; then the furnace can heat up the house.

What does all this have to do with our practice? Practice is about breaking our exclusive identification with ourselves. This process has sometimes been called purifying the mind. To "purify the mind" doesn't mean that you become holy or other than you are; it means to strip away that which keeps a person—or a furnace—from functioning best. The furnace functions best with hard coal. But unfortunately what we're full of is *soft* coal. There's a saying in the Bible: "He is like a refiner's fire." It's a common analogy, found in other religions as well. To sit through sesshin is to be in the middle of a refining fire. Eido Roshi said once, "This zendo is not a peaceful haven, but a furnace room for the combustion of our egoistic delusions." A zendo is not a place for bliss and relaxation, but a furnace room for the combustion of our egoistic delusions. What tools do we need to use? Only one. We've all heard of it, yet we use it very seldom. It's called *attention*.

Attention is the cutting, burning sword, and our practice is to use that sword as much as we can. None of us is very willing to use it; but when we do—even for a few minutes—some cutting and burning takes place. All practice aims to increase our ability to be attentive, not just in zazen but in every moment of our life. As we sit we grasp that our conceptual thought process is a fantasy; and the more we grasp this the more our ability to pay attention to reality increases. One of the great Chinese masters, Huang Po, said: "If you can only rid yourselves of conceptual thought, you will have accomplished everything. But if you students of the Way do not rid yourselves of conceptual thought in a flash, even though you strive for eon after eon, you will never accomplish it." We "rid ourselves of conceptual thought" when, by persistent observation, we recognize the unreality of our self-centered thoughts. Then we can remain dispassionate and fundamentally unaffected by them. That does not mean to be a cold person. Rather, it means not to be caught and dragged around by circumstances.

Most of us are not much like this. As soon as we get into our work day, we discover we're not at all calm. We have many emotional opinions and judgments about everything; our feelings are easily hurt. We're by no means "dispassionate and fundamentally unaffected" by what is going on. So it's extremely important to remember that the main purpose of doing sesshin is this burning out of thoughts by the fire of attention, so that our lives *can be* dispassionate and fundamentally unaffected by outward circumstances. I don't think there's anyone here of whom that is wholly true. Yet our practice is to do that. If we truly accomplished this burning out of attachments there would be no need to sit. But I don't think anyone can say that. We need an adequate daily period of zazen in which we attend to what's going on in our minds and bodies. If we don't sit regularly, then we can't comprehend that how we wash our car or how we deal with our supervisor is absolutely our practice.

Master Rinzai said, "We cannot solve past karma except in relationship to circumstances. When it is time to dress, let us put on our clothes. When we should take a walk, let us walk. Do not have a single thought in mind about searching for Buddhahood." Somebody once asked me, "Joko, do you think you're ever going to achieve great and final enlightenment?" I replied, "I hope a thought like that would ever occur to me." There is no special time or place for great realization. As Master Huang Po said, "On no account make a distinction between the Absolute and the sentient world." It's nothing more than parking your car, putting on your clothes, taking a walk. But if soft coal is what we're burning, we're not going to realize that. Soft coal simply means that the burning in our life is not clean. We are unable to burn up each circumstance as we encounter it. And the culprit is always our emotional attachment to the circumstance. For example, perhaps your boss asks you to do something unreasonable. At that moment what is the difference between burning soft coal and hard coal? Or suppose we are looking for employment—but the only work we can find is something we dislike. Or our child gets into trouble at school . . . In dealing with those, what is the difference between soft coal and hard coal? If there isn't some comprehension of the difference, we have wasted our hours in sesshin. Most of us are here chasing after Buddhahood. Yet Buddhahood *is* how you deal with your boss or your child, your lover or your partner, whoever. Our life is always absolute: that's all there is. The truth is not somewhere else. But we have minds that are trying to burn the past or the future. The living present—Buddhahood—is rarely encountered.

When the fire in the furnace is banked, and you want a brightly burning fire, what do you do? You increase the air intake. We are fires too; and when the mind quiets down we can breathe more deeply and the oxygen intake goes up. We burn with a cleaner flame, and our action comes out of that flame. Instead of trying to figure out in our minds what action to take, we only need to purify the base of ourselves; the action will flow out of that. The mind quiets down because we observe it instead of getting lost in it. Then the breathing deepens and, when the fire really burns, there's nothing it can't consume. When the fire gets hot enough, there is no self, because now the fire is consuming everything; there is no separation between self and other.

We don't like to think of ourselves as just physical beings; yet the whole transformation of sitting is physical. It's not some miraculous thing that happens in our head. When we burn soft coal, we are misusing our minds so that they are constantly clogged with fantasies, opinions, desires, speculations, analysis—and we try to find right action out of that bog. When something goes wrong in our life, what do we try to do? We sit down, try to figure it out, mull it over, speculate about it. That doesn't work. What does work is noticing our mental aberrations—which are not true thinking. We observe our emotional thoughts. "Yeah, I really can't stand her! She's a terrible person!" We just notice, notice, notice. Then, as mind and body quiet down and the fire burns brighter, out of that will come real thinking and the ability to make adequate decisions. The creative spark of any art is also born in that fire.

We want to think. We want to speculate. We want to fantasize. We want to figure it all out. We want to know the secrets of the universe. When we do all that, the fire stays banked; it's not getting any oxygen. Then we wonder why we're sick, mentally and physically. The burning is so clogged, there's nothing but debris coming off. And that debris doesn't just dirty us; it dirties everything. So it's important to sit every day; otherwise the understanding of the burning process gets so dim and cloudy that the fires stay banked. We have to sit every day. Even ten minutes is better than not sitting at all. Sesshins are also essential for serious students; daily sitting may keep a low-grade fire burning, but usually it doesn't burst into a full blaze.

So let's just continue with sesshin. There's nothing you won't face before you're done with it: rage, jealousy, bliss, boredom. Watch yourself as you cling to feeling sorry for yourself; as you cling to your problems, as you cling to the "awful" state of your life. That's your drama. The truth is, we like our drama very much. People tell me they want to be free of their troubles; but when we stew in our own juices we can maintain ourselves as the artificial center of the universe. We love our drama. We like to complain and agonize and moan. "Isn't it terrible! I'm so lonely! Nobody loves me." We enjoy our soft coal. But the messiness of that incomplete burning can be tragic for me and for you. Let's practice well.

Pushing for Enlightenment Experiences

One of my favorite lines in the *Shōyō Rōku* says, "On the withered tree, a flower blooms." When all human grasping and human need are ended, there is wisdom and compassion. This is the state of a Buddha. Personally I doubt that there ever was a person who completely realized this state. Or perhaps there have been a few in the history of humankind. But we confuse people who have great power and insight with the reality of a completely enlightened Buddha. So let's look at what the process of becoming a Buddha might be, working backwards.

For this fully enlightened (and perhaps hypothetical) being, there would be no boundaries. There would be nothing in the universe about which such a being could not say without qualification, *Namu Dai Bosa*, "Unite with Great Enlightened Being." You and I cannot say this truly for everything. All we can do is to extend our ability to do so. But a Buddha would be one who could say that, who could be united without barrier or boundaries with everything in the universe.

Now before such complete enlightenment there's a state of a fully integrated person. Of course for this person there are still boundaries, limitations, so there is some place where that integration fails. Nonetheless that's what you might call mind/body integration, wonderful and rare. Most of us are in some of the stages leading up to that, which means we cannot own even our own bodies completely. Any tension in the body means that we cannot own it. We won't say that we *are* a body, but that we *have* a body. And then there's a state before that, when we completely disown the body, thinking we're just a mind. And there's a state before that in which we cannot even own all of our mind; we split some of it off as well.

Depending on what our conditioning is right now, we can see just so much, and we can embrace just so much. The last state I mentioned is so constricted, so narrow, that anything introduced beyond it is fearsome. If introduced too soon it's devastating. And this is where we encounter many of the odd and harmful effects of a practice. For this constricted person the universe looks like a little pinpoint of light. Introduce a light as bright as the sun and that person may go crazy, and sometimes does.

I've been at sesshins where there's screaming, yelling, pushing: you've got to *do it!* You've got to *die!* The women weep all night, the men weep all night, and for a few people who are ready for this amount of pressure, that's fine. Some people who are not ready, and who are good girls and boys, will concentrate and cut through, bypassing all those early stages of development to a point where for a moment they see, they have an "opening." Is that good? No, not necessarily. For those who are ready, that expe-

rience is the most wonderful thing in the world. They sense it
before they have it and they're prepared to receive it. But for someone who is not prepared, it can be harmful. It produces no good
results: in fact quite the opposite may be true.

A teacher may deliberately narrow and concentrate a student's
vision by instructing the student to work on a koan like Mu.* But
a person who is not emotionally ready for such an endeavor might
do better to practice in a different way. Great care must be exercised; a premature enlightenment experience is not necessarily
good. To have such an experience is to realize that we are nothing
(no-self), and that there is nothing in the universe but change. We
encounter this enormous elemental power which we are. To realize this when ready is liberating. But for a person who's not ready,
it's annihilation. And even a person who is ready for such an
experience may have to spend many years practicing with the
bypassed levels of maturation, clearing them up.

Some teachers have had enormous experience with advanced
states, but not with the earlier levels. Sure they see. But that very
vision, when not integrated firmly, can create mischief, not harmony and peace.

We may believe that an enlightenment experience is like having
a piece of birthday cake. "Exciting! I want to have that!" But someone has spoken of this experience as being a terrible jewel. Unless
the structure is firm enough to support it, the whole structure
may collapse. It's not wise to take just anyone off the street and
push them. Some teachers don't understand that: they work
intuitively, but without enough understanding of the differences
in people. Years ago I asked a great pianist, "How can I improve
the way I play this passage? I'm having difficulty with it." And she
replied, "Oh. It's easy. Just do it like this." For her, that was clear
and easy, but for me it was of no use; the difficulty remained.

What I'm asking is that you be patient. I meet people who have
been sitting a long time, and who have power and some insight,

---

* Mu: a koan often assigned to beginning students as a means of focusing concentration. Its literal meaning—"no" or "nothing"—does not fully capture its significance in Zen practice.

but who are all screwed up because their development has not been balanced. And that balancing is not a simple thing to do. As we sit we come to know how complicated we are. And there may be various little eddies in our complicated selves where we need experts in other fields to help us. Zen will not take care of everything. When the intensity level of practice becomes too high, too soon, there's a danger of imbalance and we need to slow down. We shouldn't see too much too soon.

Why even talk about enlightenment? When a person is ready, when that urge to know is strong, it's obvious to the teacher and student what to do next. We need to work patiently with our lives, with our desires for sensation, for security, for power—and no one here is free of those, including myself. So I'm asking you to reexamine some of your thoughts about wanting to achieve enlightenment and to face this job that must be done with steadiness and intelligence. With patient practice our lives can constantly grow in power and also in integration, so that the power will be used for the good of all.

Every time we return our mind to the present that power develops. Every time we are really aware of our mental dreaming that power develops, slowly, slowly. Then there is a genuine calming and clarification of mind and body. It's obvious—we can recognize such people just by looking at them.

In this lifetime, if we practice well, there is the certainty of moving far along the path, perhaps with enlightenment experiences illuminating the way—and that's fine. But let's not underestimate the constant work we have to do on all the illusions that constantly interrupt our journey. Consider the Ox-herding Pictures,* for example: people want to jump from one to ten. But we can be at nine and slip right back to two. Advances are not always permanent and solid. We might be at ten for a few hours, and then the next day we're back at two. In retreats our minds get clear and quiet—but just let somebody come up and criticize us!

---

* The Ox-herding Pictures: a traditional series of drawings depicting the progress of practice from delusion to enlightenment, cast in the form of a man progressively taming a wild ox.

"On a withered tree, a flower blooms." Or, in the Bible, "Lest ye die, ye shall not be born again." And of course our practice is to die slowly, step by step, gradually disidentifying with wherever we're caught in. If we're caught anywhere we have not died. For example, we may identify with our family. Disidentifying with one's family doesn't mean not to love them. Or consider your husband or boyfriend or girlfriend—that need. The longer we practice, the more minimal this need becomes. The love becomes greater and the need less. We can't love something we need. If we need approval, we haven't died. If we need power, if we need to have a certain position, if it's not okay with us to do the most menial job, we haven't died. If we need to be seen in a particular way, we haven't died. If we want to have things our way, we haven't died. I haven't died in any of these ways. I'm just very aware of my attachments and I don't act on them very often. But having died means they're not there. In this sense a truly enlightened being is not human—and I don't know anyone like that. I've been around some remarkable persons during a lifetime and still I haven't met anyone like that. So let's be content to be where we are and working hard. For us to be as we are at this point in time is perfect.

As we identify ourselves with less and less, we can include more and more in our lives. And this is the vow of the boddhisattva. So the degree to which our practice ripens, to that degree we can do more, we can include more, we can serve more; and that's really what Zen practice is about. Sitting like this is the way; so let's just practice with everything we have. All I can be is who I am right now; I can experience that and work with it. That's all I can do. The rest is the dream of the ego.

The Price of Practice

When we find our life unpleasant or unfulfilling, we try to escape the unpleasantness by various subtle escape mechanisms. In such attempts we are dealing with our lives as if there's *me* and then there's *life outside me*. As long as we approach our lives in this way we will bend all of our efforts to finding something or somebody else to handle our lives for us. We may look for a lover, a teacher, a religion, a center—something, or somebody, somewhere, to handle our difficulties for us. As long as we see our lives in this dualistic fashion we fool ourselves and believe that we need not pay any price for a realized life. All of us share this delusion to varying degrees; and it leads only to misery in our lives.

As our practice proceeds the delusion comes under attack; and slowly we begin to sense (horror of horrors!) that *we* must pay the price of freedom. No one but ourselves can ever pay it for us. When I realized that truth it was one of the strong shocks of my lifetime. I finally understood one day that only *I* can pay the price of realization: no one, no one at all, can do this for me. Until we understand that hard truth, we will continue to resist practice; and even after we see it our resistance will continue, though not as much. It is hard to maintain the knowledge in its full power.

What are some of the ways in which we evade paying the price? The chief one is our constant unwillingness to bear our own suffering. We think we can evade it or ignore it or think it away, or persuade someone else to remove it for us. We feel that we are entitled not to feel the pain of our lives. We fervently hope and scheme for someone else—our husband or wife, our lover, our child—to handle our pain for us. Such resistance undermines our practice: "I won't sit this morning; I just don't feel like it." "I'm not going to do sesshin; I don't like what comes up." "I won't hold my tongue when I'm angry—why should I?" We waver in our integrity when it is painful to maintain it. We give up on a relationship that no longer fulfills our dreams. Underneath all of these evasions is the belief that others should serve us; others should clean up the messes we make.

In fact, nobody—but nobody—can experience our lives for us; nobody can feel for us the pain that life inevitably brings. The price we must pay to grow is always in front of our noses; and we never have a real practice until we realize our unwillingness to pay any price at all. Sadly, as long as we evade, we shut ourselves off from the wonder of what life is and what we are. We try to hold on to people who we think can mitigate our pain for us. We try to dominate them, to keep them with us, even to fool them into taking care of our suffering. But alas, there are no free lunches, no giveaways. A jewel of great price is never a giveaway. We must earn it, with steady, unrelenting practice.

We must earn it in each moment, not just in the "spiritual side" of our life. How we keep our obligations to others, how we serve others, whether we make the effort of attention that is called for each moment of our life—all of this is paying the price for the jewel.

I'm not talking about erecting a new set of ideals of "how I should be." I'm talking about earning the integrity and wholeness of our lives by every act we do, every word we say. From the ordinary point of view, the price we must pay is enormous—though seen clearly, it is no price at all, but a privilege. As our practice grows we comprehend this privilege more and more.

In this process we discover that our own pain and others' pain are not separate worlds. It's not that, "My practice is my practice and their practice is their practice"; because when we truly open up to our own lives we open up to *all* life. The delusion of separateness diminishes as we pay the price of attentive practice. To overcome that delusion is to realize that in practice we are not only paying a high price for ourselves, but for everyone else in the world. As long as we cling to our separateness—my ideas about what I am, what you are, and what I need and want from you—that very separateness means that we are not yet paying the price for the jewel. To pay the price means that we must give what life requires must be given (not to be confused with indulgence); perhaps time, or money, or material goods—and sometimes, *not* giving such things when it is best not to. Always the practice effort is to see what life requires us to give as opposed to what we personally want to give—which is not easy. This tough practice is the payment exacted if we wish to encounter the jewel.

We cannot reduce our practice simply to the time we spend in zazen, vital though this time is. Our training – paying the price – must take place twenty-four hours a day.

As we make this effort over time, more and more we come to value the jewel that our life is. But if we continue to stew and fuss with our life as if it were a problem, or if we spend our time in seeking to escape this imaginary problem, the jewel will always remain hidden.

Though hidden, the jewel is always present – but we will never see it unless we are ready to pay the price. The uncovering of the jewel is what our life is about. How willing are you to pay the price?

The Reward of Practice

We are always trying to move our lives from unhappiness to happiness. Or we might say that we wish to move from a life of struggle to a life of joy. But these are not the same: moving from unhappiness to happiness is *not* the same as moving from struggle to joy. Some therapies seek to move us from an unhappy self to a happy self. But Zen practice (and perhaps a few other disciplines or therapies) can help us to move from an unhappy self to no-self, which is joy.

To have a "self" means we are self-centered. Being self-centered – and therefore opposing ourselves to external things – we are anxious and worried about ourselves. We bristle quickly when the external environment opposes us; we are easily upset. And being self-centered, we are often confused. This is how most of us experience our lives.

Although we are not acquainted with the opposite of a self (no-self), let's try to think what the life of no-self might be. No-self doesn't mean disappearing off the planet or not existing. It is neither being self-centered nor other-centered, but just centered. A life of no-self is centered on no particular thing, but on all things—that is, it is nonattached—so the characteristics of a self cannot appear. We are not anxious, we are not worried, we do not bristle easily, we are not easily upset, and, most of all, our lives do not have a basic tenor of confusion. And thus to be no-self is joy. Not only that; no-self, because it opposes nothing, is beneficial to everything.

For the vast majority of us, however, practice has to proceed in an orderly fashion, in a relentless dissolution of self. And the first step we must take is to move from unhappiness to happiness. Why? Because there is absolutely no way in which an unhappy person—a person disturbed by herself or himself, by others, by situations—can be the life of no-self. So the first phase of practice should be to move from unhappiness to happiness, and the early years of zazen are mostly about this movement. For some people, intelligent therapy can be useful at this point. But people differ, and we can't generalize. Nevertheless we cannot (nor should we try to) skip over this first movement from relative unhappiness to relative happiness.

Why do I say "relative" happiness? No matter how much we may feel that our life is "happy," still, if our life is based on a self, we cannot have a final resolution. Why can there not be a final resolution for a life based on a self? Because such a life is based on a false premise, the premise that we *are* a self. Without exception we all believe this—every one of us. And any practice that stops with the attempted adjustment of the self is ultimately unsatisfying.

To realize one's true nature as no-self—a Buddha—is the fruit of zazen and the path of practice. The important thing (because only it is truly satisfying) is to follow this path. As we battle with the question of our true nature—self or no-self—the whole basis of our life must change. To adequately wage this battle, the whole feeling, the whole purpose, the whole orientation of life must be transformed. What might be the steps in such a practice?

The first, as I said, is to move from relative unhappiness to relative happiness. At best this is a shaky accomplishment, one that is easily upset. But we must have some degree of relative happiness and stability to engage in serious practice. Then we can attempt the next stage: an intelligent, persistent filtering of the various characteristics of mind and body through zazen. We begin to see our patterns: we begin to see our desires, our needs, our ego drives, and we begin to realize that these patterns, these desires, these addictions are what we call the self. As our practice continues and we begin to understand the emptiness and impermanence of these patterns, we find we can abandon them. We don't have to *try* to abandon them, they just slowly wither away—for when the light of awareness plays on anything, it diminishes the false and encourages the true—and nothing brightens that light as much as intelligent zazen, done daily and in sesshin. With the withering of some of these patterns, no-self—which is always present—can begin to show itself, with an accompanying increase of peace and joy.

This process, though easy to talk about, is sometimes frightening, dismal, discouraging; all that we have thought was ourself for many years is under attack. We can feel tremendous fear as this turning about takes place. It may sound enchanting as it is talked about, but the actual "doing" can be horrendous.

Still, for those of us who are patient and determined in our practice, joy increases; peace increases; the ability to live a beneficial and compassionate life increases. And the life which can be hurt by the whims of outside circumstances subtly alters. This slowly transforming life is not, however, a life of no problems. They will be there. For a time our life may feel worse than before, as what we have concealed becomes clear. But even as this occurs, we have a sense of growing sanity and understanding, of basic satisfaction.

To continue practice through severe difficulties we must have patience, persistence, and courage. Why? Because our usual mode of living—one of seeking happiness, battling to fulfill desires, struggling to avoid mental and physical pain—is always undermined by determined practice. We learn in our guts, not just in our brain, that a life of joy is not in seeking happiness, but in experiencing and simply *being* the circumstances of our life as they are; not in fulfilling personal wants, but in fulfilling the needs of life; not in avoiding pain, but in being pain when it is necessary to do so. Too large an order? Too hard? On the contrary, it is the easy way.

Since we can only live our lives through our minds and bodies, there is no one who is not a psychological being. We have thoughts, we have hopes, we can be hurt, we can be upset. But the real solution must come from a dimension which is radically different from the psychological one. The practice of nonattachment, the growth of no-self, is the key to understanding. Finally we realize that there is no path, no way, no solution; because from the beginning our nature *is* the path, right here and right now. Because there is no path our practice is to follow this no-path endlessly—and for no reward. Because no-self is everything it needs no reward: from the no-beginning it is itself complete fulfillment.

III. FEELINGS

III. FEELINGS

A Bigger Container

At the age of ninety-five Genpo Roshi, one of the great Zen masters of modern times, was speaking of the "gateless gate," and he pointed out that there truly is *no* gate through which we must pass in order to realize what our life is. Still, he said, from the standpoint of practice we must go through a gate, the gate of our own pride. And everyone of us here, since the time we got up this morning, has in some way or another met our pride—every one of us. To go through this gate that is not a gate we have to go through the gate of our own pride.

Now the child of pride is anger. By anger I mean all kinds of frustration, including irritation, resentment, jealousy. I talk so much about anger and how to work with it because to understand how to practice with anger is to understand how to approach the "gateless gate."

In daily life we know what it means to stand back from a problem. For example, I've watched Laura make a beautiful flower arrangement: she'll fuss and fiddle with the flowers, then at some point she'll stand back and look, to see what she has done and how it balances out. If you're sewing a dress, at first you cut and arrange and sew, but finally you get in front of the mirror to see how it looks. Does it hang on the shoulders? How's the hem? Is it becoming? Is it a suitable dress? You stand back. Likewise, in order to put our lives into perspective, we stand back and take a look.

Now Zen practice is to do this. It develops the ability to stand back and look. Let's take a practical example, a quarrel. The overriding quality in any quarrel is pride. Suppose I'm married and I have a quarrel with my husband. He's done something that I don't like—perhaps he has spent the family savings on a new car—and I think our present car is fine. And I think—in fact I *know*—that I am right. I am angry, furious. I want to scream. Now what can I do with my anger? What is the fruitful thing to do? First of all I think it's a good idea just to back away: to do and say as little as possible. As I retreat for a bit, I can remind myself that what I really want is to be what might be called A Bigger Container. (In other words I must practice my ABCs.) To do this is to step into another dimension—the spiritual dimension, if we must give it a name.

Let's look at a series of practice steps, realizing that in the heat of anger it's impossible for most of us to practice as the drama occurs. But do try to step back; do and say very little; remove yourself. Then, when you're alone, just sit and observe. What do I mean by "observe"? Observe the soap opera going on in the mind: what he said, what he did, what I have to say about all that, what I should do about it . . . these are all a fantasy. They are not the reality of what's happening. If we can (it's difficult to do when angry), label these thoughts. Why is it difficult? When we're angry there's a huge block that stands in the way of practice: the fact that we don't *want* to practice—we prefer to cherish our pride, to be "right" about the argument, the issue. ("Do not seek the Truth—only cease to cherish opinions.") And that's why the first step is to back away, say little. It may take weeks of hard practice before we can see that what we want is not to be *right*, but to be A Bigger Container, ABC. Step back and observe. Label the thoughts of the drama: yes, he shouldn't do that; yes, I can't stand what he's doing; yes, I'll find some way to get even—all of which may be so on a superficial level, but still it is just a soap opera.

If we truly step back and observe—and as I said, it's extremely difficult to do when angry—we will be capable in time of seeing our thoughts as thoughts (unreal) and not as the truth. Sometimes I've gone through this process ten, twenty, thirty times before the thoughts finally subside. When they do I am left with what? I am left with the direct experience of the physical reaction in my body, the residue, so to speak. When I directly experience this residue (as tension, contraction), since there is no duality in direct experience, I will slowly enter the dimension (samadhi) which *knows* what to do, what action to take. It knows what is the best action, not just for me but for the other as well. In making A Bigger Container, I taste "oneness" in a direct way.

We can talk about "oneness" until the cows come home. But how do we actually separate ourselves from others? How? The pride out of which anger is born is what separates us. And the solution is a practice in which we experience this separating emotion as a definite bodily state. When we do, A Bigger Container is created.

What is created, what grows, is the amount of life I can hold without it upsetting me, dominating me. At first this space is quite restricted, then it's a bit bigger, and then it's bigger still. It need never cease to grow. And the enlightened state is that enormous and compassionate space. But as long as we live we find there is a limit to our container's size and it is at that point that we must practice. And how do we know where this cut-off point is? We are at that point when we feel any degree of upset, of anger. It's no mystery at all. And the strength of our practice is how big that container gets.

As we do this practice we need to be charitable with ourselves. We need to recognize when we're unwilling to do it. No one is willing all the time. And it's not bad when we don't do it. We always do what we're ready to do.

This practice of making A Bigger Container is essentially spiritual because it is essentially nothing at all. A Bigger Container isn't a thing; awareness is not a thing; the witness is not a thing or a person. There is not somebody witnessing. Nevertheless that which can witness my mind and body must be other than my mind and body. If I can observe my mind and body in an angry state, who is this "I" who observes? It shows me that I am other than my anger, bigger than my anger, and this knowledge enables me to build A Bigger Container, to grow. So what must be increased is the ability to observe. *What* we observe is always secondary. It isn't important that we are upset; what is important is the ability to observe the upset.

As the ability grows first to observe, and second to experience, two factors simultaneously increase: wisdom, the ability to see life as it is (not the way I want it to be) and compassion, the natural action which comes from seeing life as it is. We can't have compassion for anyone or anything if our encounter with them is ensnarled in pride and anger; it's impossible. Compassion grows as we create A Bigger Container.

When we practice we're cutting deep into our life as we've known it, and the way this process unfolds varies from person to person. For some people, depending on their personal conditioning and history, this process may go smoothly, and the release is slow. For others it comes in waves, enormous emotional waves. It's like a dam that bursts. We fear being flooded and overwhelmed. It's as though we've walled off part of the ocean, and when the dam breaks the water just rejoins that which it truly is; and it's relieved because now it can flow with the currents and the vastness of the ocean.

Nevertheless I think that it's important for the process not to go too fast. If it's going too fast I think it should be slowed down. Crying, shaking, upset, are not undesirable things. That dam is beginning to break. But it's not necessary that it break too fast. Better to slow it down, and if it breaks rapidly, that also OK—it's just that it doesn't have to be done that way. We think we're all the same; but probably the more repressive and difficult the childhood has been, the more important it is for the dam to give way slowly. But no matter how smooth our life may have been, there's always a dam that has to burst at some point.

Remember also that a little humor about all this isn't a bad idea. Essentially we never get rid of anything. We don't have to get rid of all our neurotic tendencies; what we do is begin to see how funny they are, and then they're just part of the fun of life, the fun with living with other people. They're all crazy. And so are we, of course. But we never really see that we're crazy; that's our pride. Of course *I'm* not crazy—after all, I'm the teacher!

Opening Pandora's Box

The quality of our practice is always reflected in the quality of our life. If we are truly practicing there will be a difference over time. Now one of the illusions we may have about our practice is that practice will make things more comfortable, clearer, easier, more peaceful, and so on. Nothing could be further from the truth. This morning as I was drinking my coffee two fairy tales popped into my head, and I suppose that nothing pops in except for some reason or another. Fairy tales embody some basic, fundamental truths about people; that's why they have existed for as long as they have.

The first fairy tale that came into my mind was about the princess and the pea: long ago the test of a true princess was that, if she slept atop a pile of thirty mattresses, she could feel a pea beneath the bottom mattress. Now you might say that practice turns us into princesses; we become more sensitive. We know things about ourselves and others that we didn't know before. We become much more sensitive, but sometimes we become more edgy, too.

The other story is about Pandora's box. You remember — somebody was so curious about the contents of that mysterious box that he finally opened it — and the evil contents poured out, creating chaos. Practice is often like that for us; it opens Pandora's box.

All of us feel we are separate from life; we feel as if we have a wall around us. The wall may not be very visible; it may even be invisible — but the wall is there. As long as we feel separate from life we feel the presence of a wall. An enlightened person wouldn't have a wall. But I've never met someone who I felt was completely free of one. Still, with practice that wall keeps getting thinner and more transparent.

That wall has been keeping us out of touch. We may be anxious, we may have disturbing thoughts, but our wall keeps us unaware of that. But as we practice (and many of you know this very well) this wall begins to have holes in it. Before it was like a plank covering bubbling water; but now the plank has begun to develop holes, as practice makes us more aware and sensitive. We can't sit motionless for even thirty minutes without learning something. And when that thirty minutes goes on day after day after day after day, we learn and learn. Whether we like it or not, we learn.

Pieces of the plank may even fall away so that the water begins to bubble up through the holes and gaps. Of course what we have covered is that which we do not wish to know about ourselves. When it bubbles up (as it will if we practice) it's as though Pandora's box begins to open. Ideally in practice that box should never be thrown wide open all at once. But since the release is not completely predictable, there can be some surprises, even casualties. At times the lid comes off and everything we've never wanted to see about ourself comes boiling up—and instead of feeling better, we feel worse.

Pandora's box is all of our self-centered activities, and the corresponding emotions that they create. Even if we're practicing well there will be times (not for everybody, but for some people) when the box seems to explode—and suddenly a hurricane of emotions is whirling around. Most people don't like to sit when this is happening; but the people for whom this eruption resolves most easily are those who never give up sitting, whether they want to do it or not. In my own life the release went quite unobtrusively, probably because I was sitting so much and doing so many sesshins.

As practice at the Center matures I see most students' lives transforming. That doesn't mean that Pandora's box is not opening; the two go together, the transformation and the discomfort. For some this is a very painful time, when the box begins to open. For example, unexpected anger may surface (but please don't take it out on someone else). So the illusion we have, that practice should always be peaceful and loving, just isn't so. That the box opens is perfectly normal and necessary. It's not good or bad, it's just what has to happen if we really want our lives to settle down, and be more free of a reactive way of living. None of this process is undesirable; in fact, properly worked with, it's desirable. But how we practice with the boiling up is the crucial thing.

Practice is not easy. It *will* transform our life. But if we have a naive idea that this transformation can take place without a price being paid, we fool ourselves. Don't practice unless you feel there's nothing else you can do. Instead, step up your surfing or your physics or your music. If that satisfies you, do it. Don't practice unless you feel you must. It takes enormous courage to have a real practice. You have to face everything about yourself hidden in that box, including some unpleasant things you don't even want to know about.

To do Zen practice, we have to desire a certain kind of a life. In traditional terms, it's a life in which our vows override our ordinary personal considerations: we must be determined that our lives develop a universal context and that the lives of others also develop that context. If we're at a stage in our lives (and it's not good or bad, it's just a stage) in which the only thing that matters to us is how *we* feel and what *we* want, then practice will be too difficult. Perhaps we should wait a while. As a teacher I can facilitate practice and of course encourage a person's effort, but I can't give anyone that initial determination; and it has to be there for practice to take hold.

The box that right now is opening for many of you—how will you work with it? Some things you should know about this upsetting phase of practice. One, for people on this path, it's normal—in fact, necessary. Two, it doesn't last forever. And three, more than at any other time, it's a time when we need to understand our practice, and to know what patience is. And it's particularly a time to do sesshin. If you've been sitting for twenty, thirty years, whether or not you do sesshins is not as crucial. But there are certain years when it's vitally important and you should do as many as your life situation permits. And that advice presupposes the strength to maintain such intensity of practice. It's not "bad" not to want such a dedicated practice. I want to emphasize that. Sometimes people need another ten years or so of just knocking around, letting life present its lessons, before they're ready to commit to an intense practice.

So Pandora's box, that which upsets and disturbs us, is the

emergence (sometimes in a flood) of that which we have not been aware of before: our anger toward life. It has to boil out sooner or later. This is the ego, our anger that life is not the way we want it to be. "It doesn't suit me! It doesn't give me what I want! I want life to be nice to me!" It is our fury when the people or events in our lives simply don't give us what we demand.

Perhaps right now you are in the middle of opening the box. At some point I would like you to share what you have found useful in your practice at this time. A student, in some ways, can be more useful to others than a person like myself, who can hardly remember this stage. I understand the conflict pretty well, but the actual memory of how difficult it can be is fading. That's one of the great things about a sangha: it's a group of people who have a mutual framework for practice. In the sangha we can be honest, we don't have to hide or cover our struggles. The most painful thing is to think that there's something wrong with *me*, and that nobody else is having the trouble I am. That's not true, of course.

"Do Not Be Angry"

When I give a talk I'm trying to elucidate, by any means I can find, what life is about for me and what it might be for someone else, as opposed to our illusions about it. It's a very difficult thing to talk about. I never give a dharma talk that I don't hate, because it's never possible to tell the exact truth: I always tend to go a little too far this way, or a little too far that way, or I use the wrong words and somebody gets mixed up . . . but again, that's part of our training. Dharma talks are not necessarily something to understand; if they shake you up and confuse you, sometimes that's just right. For example: we can say that everybody in the universe at this particular moment is doing the best that he or she can. And then the word "best" creates trouble. It's the same difficulty we have with the sentence, "Everything, just as it is, is perfection." Perfect? Doing their best? You mean, when someone's doing something horrible they're doing their best? Just through our use of words we get awfully mixed up in our life and in our practice.

In fact our whole life is confused because we mix up our concepts (which are themselves absolutely necessary) with reality. So dharma talks tend to challenge our usual concepts. And using words in a certain way adds lots of confusion and that's just fine. Today I want to add to the confusion. I'm going to tell a little story and then head off in some other directions, and see what we make of all that. At this center we don't talk much about the precepts or the eight-fold path, for a very good reason: people misinterpret the precepts as being prohibitions, "thou shalt not's." And that's not what they are at all. Nevertheless, my talk today is about the precept "Do not be angry." I won't mention it again! But that's what the talk is about: "Do not be angry."

Suppose we are out on a lake and it's a bit foggy—not too foggy, but a bit foggy—and we're rowing along in our little boat having a good time. And then, all of a sudden, coming out of the fog, there's this other rowboat and it's heading right at us. And . . . *crash*! Well, for a second we're really angry—what is that fool doing? I just painted my boat! And here he comes—crash!—right into it. And then suddenly we notice that the rowboat is empty. What happens to our anger? Well, the anger collapses . . . I'll just have to paint my boat again, that's all. But if that rowboat that hit ours had another person in it, how would we react? You know what would happen! Now our encounters with life, with other people, with events, are like being bumped by an empty rowboat. But we don't experience life that way. We experience it as though there are people in that other rowboat and we're really getting clobbered by them. What am I talking about when I say that all of life is an encounter, a collision with an empty rowboat? What's that all about?

Let's leave that question for a moment. People often ask, "What do I get out of practice?" What is the change? What is the transformation?" Zen practice is very hard work. It's restrictive and difficult. We're told we have to sit every day. What do we get out of this? People usually think, "I'm going to improve. I'm going to get better. If I lose my temper easily, maybe after sitting I won't lose it so easily." Or, "To be truthful, I'm not so kind; maybe through sitting I'll get to be a very kind person." And this isn't quite right. So I want to tell you a few little incidents to clarify this a bit.

I want to talk about the dishpan at our house, where I live with Elizabeth. Now, since I'm retired from work, I am home most of the day. After I rinse the sink I like to set the dishpan in there like a dish, so if there's a spare cup during the day I can conceal it in the dishpan. Since that's the way I want the dishpan, it's obviously the right way, right? But when Elizabeth does the dishes, she rinses the dishpan and turns it over so it can dry. At noon I have the house to myself. But at five o'clock I know she's coming home. So I think, "Well, am I a man or a mouse? What am I going to do about the dishpan? Am I going to put it the way Elizabeth wants it?" So what do I do? Actually, I usually forget the whole thing and put it any old way.

And then there's another thing about Elizabeth. I live with her and she's wonderful. But there couldn't be two people who are more unlike. The joy of my life is to find the one item in the closet that I can throw out . . . it's great! Elizabeth has three of everything and doesn't want to throw anything out. So it means that when I want to find something, I can't find it because I've thrown it out; and when she wants to find something, she can't find it because she's got so much stuff she can't find it.

One more example and then we'll get to the point of all this: I'll tell you what it's like if I go to the movies with my daughter: "Mom, you know your taste in movies is just impossible!" And I say, "Well, you're not remembering the one we went to that *you* wanted to see! What about that?" So, squabble, squabble, squabble . . . and we end up going to a movie which may be . . . whatever it is.

What is the point of all those stories? Basically, I could care less about the dishpan. But we do not lose all our particular, little neurotic quirks from practice. Neither my daughter nor I really cares about the movie; but these little squabbling interchanges are what life is all about. That's just the *fun* of it. Do you understand? We don't have to analyze it, pick it apart, or "communicate" about it. The wonder of living with anything is . . . what? It's perfect in being as it is.

Now you may say that's all very well with things on this level, which are of course fairly trivial. What about serious problems, such as grief and anguish? What I'm saying is that they're not different. If someone close to you dies, then the wonder of life is just being that grief itself, being what you are. And being with it in the way that *you're* with it, which is your way, not my way. Practice is in just being willing to be with it as it is. Even "willing"—that word is not quite right either. Most of life, as we see it in the stories I told, is hilarious, that's all you can say about it. But we do not view it as hilarious. We think that the other person should be different: "They should be the way I think they should be." When we come to what we call "crisis points" in our life, it's not fun—I'm not saying that—but it still is as it is. It is still the perfection.

Now I want to take one more point: I think maturing practice is the ability to be with life and just be in it as it is. That doesn't mean that you don't have all your little considerations, all your stuff going on about it. You will! That's not the point. But it is *held* differently. And all of practice is to move what I call the cut-off point, so that we can hold more and more in this way. At first we can hold only certain things that way. Maybe in six months of practice you hold this much that way. Maybe in a year this. Ten years, this. And so on. But there's always that cut-off point at which you can't hold it. And we all have that point. As long as we live we're going to have that point.

As our practice becomes more sophisticated we begin to sense our tremendous deficiencies, our tremendous cruelty. We see the things in life we're not willing to take care of, the things we can't let be, the things we hate, the things we just can't stand. And if we've been practicing a long time there's grief in that. But what we fail to see is the area which with practice grows—the area in which we *can* have compassion for life, just because it is as it is. Just the wonder of Elizabeth being Elizabeth. It's not that she should possibly be different; she is perfect in being as she is. And myself. And you. Everybody. That area grows, but always there's that point where we can't possibly see the perfection, and that's the point where our practice is. If you've been sitting a short time, it's here, that's fine; why should it be anyplace else? And then over a lifetime that cut-off point just moves and it never ceases to occur. There's always that point. And that's what we're doing here. Sitting as we sit, just letting what comes up in ourselves come up, be there, and die. Come up, be there, and die. But when we get to the cut-off point we're not going to remember any of that! Because it's tough when we're at this point. Practice is not easy.

The little stuff in life doesn't bother me particularly. I *enjoy* all this little stuff that goes on. It's fun! I enjoy my squabbles with my daughter. "Mom, all these years and you still can't get a seat belt on?" "Well, *I can't.*" That's the fun, the fun of being with another person. But what about the cut-off point? That is where practice is. And to understand that and to work with it, and also remembering that most of the time we're very unwilling to work with it— that's also practice. We're not attempting to become some sort of saint, but to be real people, with all of our stuff going on and allowing it to go on in others. And when we can't do that, then we know a signal has been given: time to practice. I know—I went through a point last week. It wasn't easy. And yet, I went through it and now what awaits is the next point. It's going to come up. And it will be my practice.

As we get more sensitive to our life and what it truly is, we can't run away. We can try for awhile, and most of us will try for as long as we can. But we really can't run indefinitely. And if we've been sitting for some years it gets harder and harder to *run*. So I want you to appreciate your sitting and appreciate your life and each other. That's all this is about. Nothing fancy. And be aware of your cut-off point. It exists in all of us. You may turn away from it and refuse to see it; but if you do, you won't grow and life around you won't grow either. But probably, you can't avoid it for more than a certain length of time.

STUDENT: Sometimes when I read about Zen, it seems that you're just a spectator.

JOKO: No, no, not a spectator at all. Zen is action itself.

STUDENT: And it seems to be connected with the cut-off point. When you're at the cut-off point the action you take doesn't seem as wise as it might be. . .

JOKO: Let's return to the rowboat. For instance, most of us in dealing with young children can see that whatever they do—even if they come up and give you a kick in the shins—that's an empty rowboat, right? You just deal with it. I think the Buddha said, "All the world are my children." The point is to keep moving that cut-off point; we must practice when we can't let "all the world be my children." And I think that's what you're saying.

STUDENT: To carry that analogy one step further: say the child is not about to kick you in the shins, but is about to set fire to the house.

JOKO: Well, stop him! Take the matches away! But still he's just doing what he's doing for whatever reason. Try to find a way to help him learn from the incident.

STUDENT: But when you simply stop him, what are you doing differently than if you felt it was a personal attack?

JOKO: Well, let's face it, with our children, quite often we *do* see it as a personal attack, right? But if we think for ten seconds we usually know that we just have to deal with the behavior in a way that's appropriate for the child. And we can do this unless we feel our own ego threatened by the way our child is. And that's NOT an empty rowboat. And all parents have this reaction at times. We want our children to be perfect. They should be models because otherwise people could criticize us. And yet our children are just children. We're not perfect and they're not perfect.

STUDENT: You mentioned "*Be not angry.*" I wanted to ask you a question about that. You said that when anger comes up, let it happen; be there and let it go. But if you have an habituated anger response to something over a long period of time, how do you let it go?

JOKO: By experiencing the anger nonverbally, physically. You can't force it to go, but you don't necessarily have to visit it on other people.

STUDENT: I want to extend the rowboat analogy: If we saw that the rowboat was coming toward us and there was someone in it, we'd probably start screaming and yelling "Stop that and *keep that away*!" Whereas if it was an empty rowboat we'd probably just take our oar and push the rowboat aside, so it wouldn't crash into ours.

JOKO: Right, we'd take appropriate action.

STUDENT: I don't know about that, because often you yell anyway, even if it's an empty rowboat; you curse at the universe, or whatever!

JOKO: Yeah, it's something like the dishpan. You may yell, but there's a difference between that momentary response and thinking about it for the next ten miles.

STUDENT: But even though there isn't anyone out there, we manage to think that the universe is doing something to us. Even when it *is* an empty rowboat, we put a person there.

JOKO: Yes, right. Well, it's *always* an empty rowboat. Again, the point is, the longer we practice the less likely that is to come up. Not because we say, "I won't be angry"—the reaction just isn't there. We feel differently and we may not even know why.

STUDENT: If you do experience the anger coming up, is that a sure sign that you're at your cut-off point?

JOKO: Yes, that's why I said the title of this talk is *do not be angry*. But again, the point is to understand what practice means with anger; it's not some simple prohibition, which would be useless anyway.

STUDENT: Well, obviously I still have to practice some more. What happens to me when some kind of tragedy occurs is, "I don't deserve it," "My friend doesn't deserve it," "How can this happen?" I get all caught up in the injustice of it and start railing against the unfairness of it.

JOKO: OK. That's very difficult. Very, very difficult. Still, it's a practice opportunity.

STUDENT: I get confused when I hear about sudden enlightenment. If this is a process, how is there an enlightened state?

JOKO: I didn't say there *was*, for one thing! But an enlightenment experience—suddenly seeing reality just as it is—just means that for a moment one's personal considerations about life are gone. And for a second one sees the universal. The problem with most enlightenment experiences is that people hold on to them, treasure them, and then they become a hindrance. The point isn't the experience—it's going on with our life. And any value that experience has is within ourselves; we don't need to worry about it. For most of us that rowboat is full of other people all the time; it's very rare that it's empty. And so . . . our cut-off point is *here*, we just work where we are. Remember the two verses from the Fifth Patriarch—one is about endlessly polishing the mirror, and the other is seeing that from the very beginning there's no mirror to polish. Most people assume that since the second is the correct understanding, the first is useless. But no, our practice is paradoxically the first. Polishing that mirror. The cut-off point is where you polish the mirror. Absolutely necessary. Because only by doing that will we eventually see the perfection of everything, just as it is. We can't see that unless we go through really rigorous, stringent practice.

STUDENT: So it's good to experience anger.

JOKO: If you learn from it. I didn't say anything about putting that anger out on others. That's very different. We may do that sometimes. I'm not saying we won't; still, it's not productive to do it. The experiencing of anger is very quiet. Not anything noisy at all.

STUDENT: I think part of the problem comes when you say, "Don't be angry" and then you say, "Be angry."

JOKO: Let's be careful here . . . I'm saying that if anger is what you are, experience it. After all, it is the reality of the moment. So if we pretend anger is not there and cover it with a directive like "Do not be angry," then right away there's no chance to really know anger for what it is. The other side of anger, if we experience its emptiness and go through it, is always compassion. If we really, really go through it. OK, enough.

False Fear

Because we are all human we all tend to create a false problem. The problem exists because we have no choice but to live out of our particular and peculiar kind of mind. Our way of thinking is not the same as that of a cat or a horse or even a dolphin. Because of our misuse of our minds we become confused between two types of fear. One kind of fear is ordinary: if physically threatened we react, we do something—we may run, fight, call the police. But we do something; this is natural, ordinary fear. But most of our anxious life is not based on that, but upon false fear.

False fear exists because we misuse our minds. Because we see our self or "I" as a separate entity, we create various sentences with "I"" as the subject. Our sentences are about what has happened to this "I" or what might happen to it, or how these happenings might be analyzed or controlled—and all of this almost ceaseless mental activity entails a constant, uneasy evaluation of ourselves and others.

Because of the fear that arises from this false picture, we cannot act with any intelligence; it is a fear that attempts to maneuver and manipulate. Once we have "sized up" a situation or person we may act, but the action tends to be based on error—a falsity of thinking that there is an "I" separate from the action. We may have thoughts like this: "I may not make the grade." "I may not be impressive." "I may end up with nothing." "I'm too important to do dishes." Out of such *I*-thinking, a peculiar value system grows. We tend to value only people or events which we hope will maintain or establish a safe and secure life for this "I." We evaluate ourselves and develop various strategies to preserve the "I." In our Southern California pop psychology we may say, "I have to love myself." But who is loving whom? How on earth can an "I" manage to love a "myself"? We feel that "I have to love myself, I have to be good to myself, I have to be good to you." There is tremendous fear underlying those judgments, a fear that does not accomplish anything. We have a fictional "I" that we try to love and protect. We spend most of our life playing this futile game. "What will happen? How will it go? Will I get something out of it?" *I, I, I*—it's a mind game of illusion, and we are lost in it.

We might suppose that once we see the game, the game will be over—but no. That's like telling someone who is quite drunk not to be drunk. We are drunk, perpetually. But to bully ourselves, to exhort ourselves, does no good. "I'm not going to be like this" is not the answer. What is the answer? We have to approach the problem from another angle, to come in the back door. First, we must become aware of our illusion, our drunkenness. The old texts say, illumine the mind, give it light, be attentive. This is not the same as self-improvement, trying to fix our lives. It is shikan—just sitting, just experiencing, just knowing the illusions (the *I* sentences) for what they are.

It's not that "I" hears the birds, it's just *hearing* the birds. Let yourself be seeing, hearing, thinking. That is what sitting is. It is the false "I" that interrupts the wonder with the constant desire to think about "I." And all the while the wonder is occurring: the birds sing, the cars go by, the body sensations continue, the heart is beating—life is a second-by-second miracle, but dreaming our *I*-dreams we miss it. So let's just sit with what may seem like confusion. Just feel it, be it, appreciate it. Then we may more often see through the false dream which obscures our life. And then, what is there?

No Hope

A few days ago I was informed of the suicide of a friend of mine, a man I hadn't seen for a dozen years. Even then suicide was all he could talk about, so I wasn't surprised at the news. It's not that I think dying is a tragedy. We all die; that's not the tragedy. Maybe nothing is a tragedy, but I think we can say that to live without appreciating this life is at least a shame.

It's a precious opportunity we have, to be alive as human beings. It has been said that the chance of having a human life is something like being picked up as one grain of sand out of all the grains on the beach. It's such a rare chance and yet somehow, as in the case of my friend, some error arises. Some of that error is present in each one of us—not fully appreciating what we have just in being alive.

So today, what I want to talk about is having no hope. Sounds terrible, doesn't it? Actually, it's not terrible at all. A life lived with no hope is a peaceful, joyous, compassionate life. As long as we identify with this mind and body—and we all do—we hope for things that we think will take care of them. We hope for success. We hope for health. We hope for enlightenment. We have all sorts of things we hope for. All hope, of course, is about sizing up the past and projecting it into the future.

Anyone who sits for any length of time sees that there is no past and no future except in our mind. There is nothing but self, and self always is here, present. It's not hidden. We're racing around like mad trying to find something called self, this wonderful hidden self. Where is it hidden? We hope for something that's going to take care of this little self because we don't realize that already we are self. There's nothing around us that is not self. What are we looking for?

A student recently loaned me a book on a text by Dōgen Zenji called *Tenzo Kyōkun*. It is Dōgen Zenji's writings on his ideas of what a tenzo, a head cook, should be—the qualities of a tenzo, the life of a tenzo.

From Dōgen Zenji's point of view the tenzo should be one of the most mature and meticulous students in the monastery. If his practice is not what a tenzo's practice should be then, from Dōgen Zenji's viewpoint, the life of the entire monastery suffers. But obviously Dōgen Zenji, in describing these qualities of a tenzo and the directions for how a tenzo should do his work, is not just talking about the tenzo. He's talking about the life of any Zen student, any bodhisattva. And so it's very instructive and pertinent reading.

So what do we find as he describes this life of an enlightened tenzo? Some mystic vision? Some rapturous state? Not at all. There are many paragraphs on how to clean the sand out of the rice or the rice out of the sand. Very, very detailed. There's nothing in the management of the kitchen that Dōgen Zenji left out; he writes about where to put the ladles, how to hang the ladles, and so on.

Let me read you one paragraph. "Next, you should not carelessly throw away the water that remains after washing the rice. In olden times a cloth bag was used to filter out the water when it was thrown away. When you have finished washing the rice, put it into the cooking pot. Take special care, lest a mouse accidentally falls into it. Under no circumstances allow anyone who happens to be drifting through the kitchen to poke his fingers around or look into the pot."

What is Dōgen Zenji telling us? He didn't write this just for the tenzo. What can we all learn?

In this writing Dōgen Zenji repeats a famous story. If we understand this one story, we really understand what Zen practice is. Young Dōgen Zenji went to China to visit monasteries for practice and study. And one day at one of them, on a very hot June afternoon, he saw the elderly tenzo working hard outside the kitchen. He was spreading out mushrooms to dry on a straw mat.

He carried a bamboo stick but had no hat on his head. The sun's rays beat down so harshly that the tiles along the walk burned one's feet. [He] worked hard and was covered with sweat. I could not help but feel the work was too much of a strain for him. His back was a bow drawn taut, his long eyebrows were crane white.

I approached and asked his age. He replied that he was sixty-eight years old. Then I went on to ask him why he never used any assistants.

He answered, "Other people are not me."

"You are right," I said; "I can see that your work is the activity of the Buddha-dharma, but why are you working so hard in this scorching sun?"

He replied, "If I do not do it now, when else can I do it."

There was nothing else for me to say. As I walked on along that passage-way, I began to sense inwardly the true significance of the role of tenzo.

The elderly tenzo said, "Other people are not me." Let's look at this statement. What he is saying is, my life is absolute. No one can live it for me. No one can feel it for me. No one can serve it for me. My work, my suffering, my joy, are absolute. There's no way, for instance, you can feel the pain in my toe; or I can feel the pain in your toe. No way. You can't swallow for me. You can't sleep for me. And that is the paradox: in totally owning the pain, the joy, the responsibility of my life—if I see this point clearly—then I'm free. I have no hope, I have no need for anything else.

But *we* are usually living in vain hope for something or someone that will make *my* life easier, more pleasant. We spend most of our time trying to set life up in a way so that will be true; when, con-trariwise, the joy of our life is just in totally doing and just bearing what must be borne, in just doing what has to be done. It's not even what *has* to be done; it's there to be done so we do it.

Dōgen Zenji speaks of the self settling naturally on the self. What does he mean by that? He means that only you can experi-ence your own pain, your own joy. If there is one impression that comes into your life that is not received, then in that second you die a little bit. None of us completely lives like that, but at least we don't need to lose ninety percent of the experience of our life.

"If I do not do it now, when else can I do it." Only I can take care of the self from morning to night. Only I can receive life. And it's this contact, second by second by second, which Dōgen Zenji is talking about as he describes the day of the tenzo. Take care of this.

Take care of that, and that. Not just washing the rice but doing it carefully, grain by grain. Not just throwing the water out. Each bite we take of our food. Each word I say. Each word you say. Each encounter, each second. That's it. Not chanting with your mind somewhere else; not half doing the dishes, not half doing anything.

I can remember when I used to daydream literally four or five hours at a time. And now—sadly I see so many people dreaming their lives away. Sometimes a man or a woman dreams of an ideal partner; they dream and they dream. But when we live life in dreams and hopes, then what life *can* offer, that man or woman sitting right next to us—ordinary, unglamorous—the wonder of that life escapes us because we are hoping for something special, for some ideal. And what Dōgen Zenji is telling us is that real practice has nothing to do with that.

We're saying, once again, that zazen, sitting, *is* enlightenment. Why? Because second after second as we sit, that's it. The old tenzo spreading seaweed—that's a passionate life, spending his life preparing food for others. Actually, all of us are constantly preparing food for others. This "food" can be typing; it can be doing math or physics; it can be taking care of our children. But do we live our life with that attitude of appreciation for our work? Or are we always hoping, "Oh, somewhere there's got to be more than *this*"? Yes, we're all hoping.

Not only do we hope, but we really give our life to this hope, to these vain thoughts and fantasies. And when they don't "produce" for us, we're anxious, even desperate.

One of my students recently told me a good story. It's about a man who was sitting on his roof because a tidal wave was sweeping through his village. The water was well up to the roof when along came a rescue team in a rowboat. They tried hard to reach him and finally when they did, they shouted, "Well, come on. Get into the boat!" And he said, "No, no. God will save me." So the water rose higher and higher and he climbed higher and higher on the roof. The water was very turbulent, but still another boat managed to make its way to him. Again they begged him to get into the boat and to save himself. And again he said, "No, no, no. God will save me! I'm praying. God will save me!" Finally the water was almost over him, just his head was sticking out. Then along came a helicopter. It came down right over him and they called, "Come on. This is your last chance! Get in here!" Still he said, "No, no, no. God will save me!" Finally his head went under the water and he drowned. When he got to heaven, he complained to God, "God, why didn't you try to save me?" And God said, "I did. I sent you two rowboats and a helicopter."

We spend a lot of time looking for something called the truth. And there is no such thing, except in each second, each activity of our life. But our vain hope for a resting place somewhere makes us ignorant and unappreciative of what is here right now. So in sesshin, in zazen, what does it mean to have no hope?

It means of course to really do zazen, to just sit. Nothing is wrong with dreams and fantasies. Just don't hold on to them; see their unreality and turn away. Stay with the only thing that's real: the experiencing of breath and the body and the environment.

Now none of us wants to abandon our hope. And to be honest, none of us is going to abandon it all at once. But we can have periods when for a few minutes or a few hours, there is just what is, just this flow. And we are more in touch with the only thing we'll ever have, which is our life.

So if we practice like this, what reward will we get? If we really practice like this, it takes everything we have. What will we get out of it? The answer, of course, is nothing. So let's not have hope. We won't get anything. We'll get our life, of course, but we've got that already. So let's not be like my friend, failing to appreciate our life and our practice. This life *is* nirvana. Where did we think it was?

Let's remember that old tenzo. If we practice the way he spread seaweed, then we can be rewarded with this nothing at all.

Love

Love is a word not often mentioned in Buddhist texts. And the love (compassion) they talk about is not an emotion, at least not what we usually think of as an emotion. It's certainly not what we call "romantic" love, which has little to do with love. It's good to investigate what love is and how it is connected to practice, since the two fruits of our practice are wisdom and compassion.

Menzan Zenji (1683–1769) was one of the greatest scholars of Soto Zen and, more than some of the old masters, he makes practice clear. Sometimes we read the old texts and build up a picture of practice that has no relationship to buying bread at the grocery store. Menzan Zenji's words are plain. He says, "When, through practice, you know the reality of zazen thoroughly, the frozen blockage of emotion-thought will naturally melt away." He says, however: "If you think you have cut off illusory thought, *instead of clarifying how emotion-thought melts*, the emotion-thought will come up again, as though you have cut the stem of a blade of grass or the trunk of a tree and left the root alive." A lot of people misunderstand practice as the cutting-off of illusory thoughts. Of course thoughts are illusory but, as he says, if you cut them off instead of "clarifying how emotion-thoughts melts," you'll learn little. Many people have little enlightenment experiences; but because they have not clarified how emotion-thought melts, the sour fruits of emotion-thought will be what they eat in daily life. He writes, "Emotion-thought is the root of delusion, a stubborn attachment to a one-sided point of view, formed by our own conditioned perceptions."

Much of the practice at this Center is about clarifying how emotion-thought melts. First we have to see what it is: the emotional, self-centered thoughts that we fuss with all the time. Their absence, he states, *is* the enlightened state, satori itself. Without exception we're all caught in emotion-thought, but the degree can vary greatly. There is a vast difference between someone who's caught ninety-five percent of the time and someone who's caught five percent.

Strictly speaking, relationships are with everything—the cup, the rug, the mountains, people. But for today's talk we'll talk about relationships with people, because these always seem to produce the most difficulty. And if we haven't been hiding in a cave for twenty years, we all have a relationship with someone. In that relationship there's always some genuine love and some false love. How much of our love is genuine depends on how we practice with false love, which breeds in the emotion-thought of expectations, hopes, and conditioning. When emotion-thought is not seen as empty, we expect that our relationship should make us feel good. As long as the relationship feeds our pictures of how things are supposed to be, we think it's a great relationship.

Yet when we live closely with somebody, that sort of dream doesn't have much of a chance. As the months go by the dream collapses under pressure, and we find that we can't maintain our pretty pictures of ourselves or of our partners. Of course we'd *like* to keep the ideal picture we have of ourselves. I'd like to believe that I'm a fine mother: patient, understanding, wise. (If only my children would agree with me, it would be nice!) But still, this nonsense of emotion-thought dominates our lives.

Particularly in romantic love, emotion-thought gets really out of hand. I expect of my partner that he should fulfill my idealized picture of myself. And when he ceases to do that (as he will before long) then I say, "The honeymoon's over. What's wrong with him? He's doing all the things I can't stand." And I wonder why I am so miserable. My partner no longer suits me, he doesn't reflect my dream picture of myself, he doesn't promote my comfort and pleasure. None of that emotional demand has anything to do with love. As the pictures break down—and they always will in a close relationship—such "love" turns into hostility and arguments.

So if we're in a close relationship, from time to time we're going to be in pain, because no relationship will ever suit us completely. There's no one we will ever live with who will please us in all the ways we want to be pleased. So how can we deal with this disappointment? Always we must practice getting closer and closer to experiencing our pain, our disappointment, our shattered hopes, our broken pictures. And that experiencing is ultimately nonverbal. We must observe the thought content until it is neutral enough that we can enter the direct and nonverbal experience of the disappointment and suffering. When we experience the suffering directly, the melting of the false emotion can begin, and true compassion can emerge.

Fulfilling our vows is the only thing we can do for one another. The more we practice over the years, the more an open and loving mind develops. When that development is complete (which means that there is nothing on the face of the earth that we judge), that is the enlightened and compassionate state. The price we must pay for it is lifelong practice with our attachment to emotion-thought, the barrier to love and compassion.

IV. RELATIONSHIPS

The Search

Every moment of our life is relationship. There is nothing except relationship. At this moment my relationship is to the rug, to the room, to my own body, to the sound of my voice. There is nothing except my being in relationship at each second. And as we practice what grows in our life is this: first, our realization that there is nothing but being in relationship to whatever is happening in each moment; and second, our growing commitment to this relationship. Now that seems simple enough—so what interferes? What blocks our commitment to a specific human relationship, or to studying, to working, to having a good time? What is there that blocks relationship?

Because we don't always understand what it means to be in relationship to the present moment, we search. When I take phone calls coming into the Center I ask, "Well, what is on your mind?" And they may say, "I'm a person who's searching." They are saying they seek a spiritual life. People new to the center will tell me, "I'm here because I'm searching." That's fine as an initial orientation to practice; we will search if we sense that our life is lacking something. In traditional terms we're looking for God; or in modern terms for "my true self," "my true life," whatever. But it's very important to understand what it means to be searching. If we want a life of sanity, or clarity, or peace, we have to understand what all this searching is about.

What are we searching for? Depending on our particular life, our background and conditioning, what we search for may seem different from one person to another; but really we're all looking for an ideal life. We may define it as an ideal partner, an ideal job, and ideal place to live. Even if other people's ideals sound very foreign to us, people are sure of what it is they think they have to find. And they are searching for it.

In a practice like this we tend to search for something called the "enlightened" state. That's a subtle form of this search. But you have to know where to look. If you look in the skies of San Diego at night, hoping to see the Southern Cross, you'll never find it. All you have to do is go to Australia, and there it is. We have to know what it means to look. We have to turn around our ideas about this searching. And practice is a kind of turning around. Enlightenment isn't something we can search for, but we think we must search for something. So what are we doing?

Although I am at the center of my life, being in this center doesn't interest me. Something seems to be missing right here, so I'm interested in searching for the missing part. I wander out from the center, as the rays of a wheel do. First I go out here, then I go out there. I try this, I reject that. This looks favorable. That doesn't. I'm searching, searching, searching. Perhaps I'm looking for the right partner: "Well, she has some qualities, but she certainly falls short in some areas." Depending how uneasy we are, we search, and we search, and we search. We may feel we never have the right job. So we search and fuss. We're either going to improve the job we have, or we think, "I may not tell anybody, but I'm not going to be here that long!" And in a sense, that's OK. I'm not saying to stay forever with a particular job. It's not the impatient action that is invalid, it's the fact that we think the searching itself is valid.

If we cease looking, searching, what are we left with? We're left with what's been right there at the center all the time. Underneath all that searching there is distress. There is unease. The minute that we realize that, we see that the point isn't the search, but rather the distress and unease which motivate the search. That's the magic moment—when we realize that searching outside of ourselves is not the way. At first it dawns on us just a little bit. And its gets clearer over time, as we continue to suffer. See, anything that we search for is going to disappoint us. Because there are no perfect beings, perfect jobs, perfect places to live. So the search ends exactly in one place, which is . . . disappointment. A good place.

If we have any brains at all, it finally dawns on us: "I've done this before." And we begin to see that it isn't the searching that's at fault, but something about where we look. And we return more and more to the disappointment, which is always at the center. What's underneath all that search *is what*? Fear. Unease. Distress. Feeling miserable. We're in pain and we use the search to alleviate that pain. We begin to see that the pain comes because we are pinching ourselves. And just this knowledge is relief, even peace. The very peace we've been searching for so hard lies in recognizing this fact: I'm pinching myself. No one's doing it to me.

So the whole search begins to be abandoned and instead of searching, we begin to see that practice isn't a search. Practice is to be with that which motivates the search, which is unease, distress. And this is the turning around.

It never happens all at once. Our drive to go after things is so powerful it overwhelms us. No matter what I say, after we all leave here, in five minutes we'll all be looking around for something to save us. As the vow says, "Desires are inexhaustible." But you won't exhaust desires by searching; you will exhaust them by experiencing that which underlies them.

And so we begin to have an understanding of the necessity for practice. Practice isn't something we take up, like swimming lessons. People say to me, "I don't have time for my practice this semester, Joko, I'm too busy. When I have more time, I'll get back to my practice." That shows no understanding of what practice is. Practice *is* being too busy, being harried—just experiencing that.

So there are two questions. The first is, do I understand the necessity for practice? And by that I do not mean just sitting zazen. Do I understand the necessity for my whole life to be practice? And the second is, do I know what practice is? Do I really know? I've met people who have been doing something for twenty years that they called practice. They could better have been working on their golf stroke.

So right now each one of us can look at our own life. What are we searching for? If we begin to see through the searching, do we see where we must look? Do we see what we can do? A willingness to practice will come out of the conviction that there's nothing else to be done. And that decision may take twenty-five years. So there are two questions: Do I understand the necessity for practice, and do I know what practice is?

STUDENT: I think practice is moment to moment being open to all the sensory input that's coming in to me, as well as my thoughts.

JOKO: Experientially that's true, though it needs to be expanded. But in terms of how we practice, that's fine.

STUDENT: I think practice is being aware of the distress and unease that lie within, and working with that in our relationships.

JOKO: What does it mean to work with it?

STUDENT: When we're really angry, for instance: to be the anger, to experience it physically, to see the thoughts it generates.

JOKO: Yes—though sometimes people tell me they're doing this, and it's obvious they're not.

STUDENT: It's because we're not getting in there and really allowing ourselves to feel and experience that particular distress at that particular time.

JOKO: I agree. But suppose now that you're giving an introductory workshop. If you said both of these things, people would look at you and say "Huh? What are you talking about?" Or people might say, "Well, I'm being my anger but nothing happens." It's not so easy to understand the words.

STUDENT: Practice is learning to be totally with the moment, with whatever we call "now." It's learning to be here, to be now.

JOKO: The problem is that most of us interpret the moment in some pretty way. "Learning to be with the moment" sounds great. But if somebody's just told me, "That presentation you just gave was awful, Joko," I don't want to be with that moment. No one wants to experience humiliation.

STUDENT: It seems that if I am really being my anger, I could get very angry and in that direct experience, I could kill someone.

JOKO: No, if we really experience anger, we won't do that. If we believe our angry *thoughts*, we might hurt somebody. But pure experience has no verbal component, so there would be nothing to do. Pure anger is very quiet. And you would never hurt anyone with it.

Practice doesn't mean, in the middle of a fight with somebody, to stop and say, "I'm going to experience this." The more mature our practice, the more we can do that naturally as the anger arises. But most people, when they get angry, act out of their thoughts; and so they nearly always have to return later and go into the experience of upset because they're not skilled enough to do that at the time they feel threatened.

STUDENT: Practice has something to do with attention. When I totally turn my attention to something—say a situation with my son—something dynamic happens, not from my personality or good ideas.

JOKO: Yes, that's true, because there's no dualism. In complete experience there isn't me having the experience, there's just the experience. And when there's no separation then there's power and there is also a knowledge of what to do. As you say, something dynamic happens. But it's not so often that we really experience anything. We know all the verbiage, but we rarely get around to it because it's painful.

STUDENT: Part of my searching at this time involves being willing to stay in uncomfortable situations or with unpleasant feelings within myself, in an effort to be more intimate with the blind spots that obscure the moment.

JOKO: That's right, as long as it's not just an idea.

STUDENT: Usually, it is!

JOKO: Yes, with most of us, usually it is. We can talk a blue streak after a while, which is one reason why so-called advanced students are always the difficult ones. They think they know. And they don't know. They're just talking.

STUDENT: The words that come to my mind about practice are "vulnerability" and "living with." It's that effort to function without the self-protection mechanisms operating, or at least to be aware of them.

JOKO: That's true. However, with most of us self-protection is automatic. That's where anger comes from. What would be another way of talking about vulnerability?

STUDENT: You haven't closed the door on your feelings.

JOKO: Vulnerability means I won't shut the door even though I'm being hurt. The reason I want to leave the door open is so that if I feel pain, I can get out. The whole point is that I may feel pain, but I'm not going to leave just for that reason. I often notice that when people get up from the table on the patio, they don't push their chair back in. They have no commitment to that chair. They feel, "The chair isn't important, I have to get into the zendo and hear about the *truth*." But the truth *is* the chair. It's where we are right now. When we leave the door open, it's that part of us that does not want to be in relationship to anything, so we run out the door. We're looking for the truth instead of being the unease and distress of where we are right now.

Practicing with Relationships

The mind of the past is ungraspable;
the mind of the future is ungraspable;
the mind of the present is ungraspable.
DIAMOND SUTRA

What is time? Is there time? What can we say about our daily life in connection with time, and with no-time, no-self? What can we learn about relationships in connection with this no-time, no-self?

Ordinarily we think of a dharma talk or a concert or any event in life as having a beginning, a middle, and an end. But at any point in this talk, for instance, if I stop right now, where are the words I've already said? They just don't exist. If I stop at any later point in the talk, where are the words that have been said up to that point? They don't exist. And when the talk's over, where is the talk? There is no talk. All that's left are memory traces in our brains. And this memory, whatever it is, is fragmentary and incomplete; we remember only parts of any actual experience. The same thing could be said for a concert—in fact we can say the same thing about our whole day, and our whole life. At this very point in time, where is our past life? It doesn't exist.

Now, how does this pertain to relationships, to our relationships with anything and anyone—to our relationship to our sitting cushion, to our breakfast, to a person, to the office, to our children?

The way we usually hold a relationship is that, "This relationship is there, *out* there, and it's supposed to give *me* pleasure. At the very least, it shouldn't give me discomfort." In other words we make this relationship into a dish of ice cream. That dish of ice cream is there to give me pleasure and give me comfort. And very few of us view our relationship in any other light than, "There it is; I've picked you out, and you know what you're supposed to do." So ordinarily when we worry about relationships, we're not talking about the nice parts. Often the nice parts may even be predominant. But what we're interested in is the *unpleasantness*: "It shouldn't be there." And when I say "unpleasant," it could range from just annoyance to a state more intense than that.

So how is all this related to no-time, no-self?

Let's take a quarrel at breakfast. At lunchtime we're still upset; not only upset but we're telling everybody about it, getting comfort, sympathy, agreement—and already we're in our heads. "When I see him tonight we'll really have to discuss it; we'll have to really get at this matter." So there's the breakfast quarrel, there's the luncheon upset, and then there's the future—what we're going to do about the upset.

But what's really *here*? What's really *now*? As we sit having our lunch, where is that breakfast quarrel? Where is it? "The mind of the past is ungraspable." Where *is* it? The dinner, when we're going to really fix all this up (to our satisfaction, of course), where *is* it? "The mind of the future is ungraspable." It doesn't exist.

What *does* exist? What's real? There is just my upset right now, at lunch. My story describing what happened at breakfast is not what happened. It's *my story*. What is real is the headache, the fluttering in my tummy. And my chattering is a manifestation of that physical energy. Outside of the physical experience, there is nothing else that's real. And I don't know if *that's* real, but that's all we can say about it.

A few weeks ago a young woman (not a student of Zen) came to talk to me and wanted to tell me about what her husband did to her three weeks before. She was very, very upset; she could hardly speak she was so upset. So I said, "Where is your husband right now?" "Oh, my husband's at work." "Well, where is the upset, where is this quarrel, where is it?" "Well, I'm telling you about it." I said, "But where is it? Show it to me." "Well I can't show it to you, but I'm telling you about it. See, this is the way it was." "But when was it?" "Three weeks ago." "Where is it?" "Oh . . ." She was getting more and more annoyed. Finally she could see that none of the upset had any reality whatsoever. And then she said, "But if that's all there is, how can I fix up my husband?"

Now the point is that we build up elaborate systems, emotions, and drama out of our belief in time—past, present, and future. Every one of us has done this. And believe me, doing this is no trivial thing. People have put themselves—and I've done it too— into such a state they can hardly function; they can't take care of their obligations, and they make themselves sick, physically and mentally.

Now, does this mean that we do nothing if we're upset? No, we do what we do. Definitely we just do what we do and at every point we are doing the best we can.

But action based on confusion and ignorance leads directly to more confusion, upset, and ignorance. It's not good or bad, and we all do it without exception. So in our ignorance, in our belief in this linear life—"That happened yesterday," and "Here it is and it's going to go on and on and on"—we live in a world of complaints, as a victim or an aggressor, in what seems to be a hostile world.

Now just one thing and one thing alone creates this hostile world, and that is our thoughts—our pictures and our fantasies. They create a world of time and space and suffering. And yet, if we try to find the past and the future that our thoughts dwell upon, we find it is impossible—they are ungraspable.

One student told me he had been climbing a wall since he heard me talk about time, because he's been looking for his past. He said, "If there is no past and no future and I can't even get hold of the present—I mean I try to get hold of it, and it's gone—then who am I?" A good question; one that we can all ask. "Who am I?"

Let's take a typical thought, the sort we all have: "Bill makes me sick." Already there is me and Bill and this feeling sick, this emotion. There's me and Bill and the sickness. Everything is all spread out. Right now I've created me, I've created Bill, and somehow, out of all that, there's this upset.

Now let's say it instead: "Me/Bill/sick." All one. "MeBillsick." Just the experience, as it is, right now. And always we'll find that if we just are the experience, the solution is contained in it. And not even just contained in it; the experience *itself* and the solution are not two separate things. But the minute we say, "She made me sick." "He annoys me." "We did this." "She did that." "It makes me sick; makes me annoyed; really hurts my feelings," then we have you, the other person, and whatever you're cooking up about it. Instead of: there's nothing—except this very present ungraspable moment of meyouanger. Just being that, right there the solution is obvious.

But as long as we spin our thoughts, such as, "Bill makes me sick," we have a problem. You'll notice that the sentence has a beginning, a middle, and an end; and out of that comes this world: hostile, frightening, and separate.

See, there is nothing wrong with our sentences. And we all have to live in a relative world; it looks like breakfast, lunch, and dinner.

And there's nothing wrong with the conceptual relative world. What goes "wrong" is that we don't see it for what it is. And not seeing it for what it is, we tend to pick our friends and lovers much as we turn on the TV.

For instance, we meet a nice girl and, "Hm, she looks like Channel 4 and I'm always calm and comfortable with Channel 4; I know what to expect on Channel 4; a certain range of this and that; a little news—I can be pretty comfortable with a Channel 4 person." So we get together and for a while everything goes very well. There is a lot of comfort and agreement. It seems like a great relationship.

But lo and behold, what happens after a while? Somehow Channel 4 has switched over to Channel 63, a lot of irritation and anger; sometimes Channel 49, all dreams and fantasies. And what am I doing during all this? See, I was pretending to be just a Channel 4 person. But no, it seems I like to spend a lot of time at Channel 33 with childhood cartoons, mostly about my dream prince or princess. And then I have other channels like Channel 19—gloom, depression, and withdrawal. And sometimes, just when I'm into gloom, depression and withdrawal, she's into fantasy and light; that doesn't fit very well. Or sometimes all the channels seem to play at once. We have upset and a lot of noise, and one or both of the partners fights or withdraws.

What to do? We are now into our usual mess, our usual scenario; and we have to try to fix it, don't we? Somehow, all was happy once. So what we've got to do, *obviously*, is to make both of us get back on Channel 4. And we say to her, "You should be like this; you should do that: that's the person I fell in love with." For a while both parties make an effort, because there is an artificial peace on Channel 4 (and a lot of boredom). Actually most marriages look like this after a time. Somebody said you can tell who in a restaurant is married—it's the couple who don't talk to each other.

It's interesting that the question nobody asks as the channels become confused is, "*Who* turned the channels on? *Who* is the *source* of all this activity?" In a way there is nothing wrong with the channels. But we never ask who turned the channels on. Who turned our acts on? What's the source? This is the key question to ask.

If we don't ask this question, and the suffering gets bad enough, sometimes we just leave the relationship and look for a new Channel 4—because if we like Channel 4s we tend to keep picking them. And this whole scenario is true not only for intimate relationships, but at the office, on a vacation, or anywhere. This is what we do.

After a number of these unfortunate episodes, we may begin to look at the whole picture of our life. Once in a while, a rare, lucky individual really begins to examine this whole question of what he's doing with his life, and begins to ask the basic questions, "Who am I? Where did I come from? Where will I go?"

Sometimes, very sadly, we may realize that after living with someone for a long time, we have never met him, have never known him. I did this for fifteen years. Some people live out a lifetime and never meet. Their channels meet once in a while, but *they* never meet.

Then we may be fortunate and encounter a great teaching. And in the Buddhist tradition the Buddha's teaching says, "*It* completely clears all pain. This is the truth, not a lie." We may not have any idea what this means; but, if we are among the lucky ones, we may begin an intelligent practice in an effort to understand the teaching.

Intelligent zazen means making a subtle shift *constantly*, step by step; first from the grosser levels to the more subtle, and to the more subtle, and to the more subtle; beginning to see right through what we call our personality, this one that we've been talking about. We begin to really look at the mind, the body, the thoughts, the sense perceptions, everything that we thought was ourself.

The first part of our practice is as if we were in the middle of a confused, busy street; we can hardly find an empty place and the traffic is going every which way. It's confusing and frightening. And that's the way our life feels to most of us. We're so busy jumping out of the way of what's coming toward us that we can't understand our own entrapment in the traffic. But if we watch it for a while we begin to see that there are holes in the traffic here and there. We might even step up on the sidewalk and begin to take a more objective look. And no matter how busy the traffic, here and there, we begin to notice clear areas.

Now our third step might be to go into a tall building and climb up onto the third-floor balcony and observe the traffic from there. Now it looks different; we can see the direction of it, which way it's moving. We see that in a way it doesn't have anything to do with us, it's just going on.

If we climb higher and higher and higher, eventually we see that the traffic is just patterns; it's beautiful, not frightening. It's just what it is and we begin to see it as a tremendous panorama. We begin to see areas of difficulty as part of the whole, not necessarily good or bad; just part of life. And after years of practice we may reach a place where we just enjoy what we see; enjoy ourselves, enjoy everything just as it is. We can enjoy it but not be caught by it, seeing its impermanence, its flow.

Then we go further, to the stage of being the witness of our life. It's all going on, it's all enjoyable; we're not caught by any of it. And in the final state of our practice we're back in the street, back in the marketplace, right in the middle of the hubbub. But seeing the confusion for what it is, we're free of it. We can love it, enjoy it, serve it, and our life is seen as what it always has been—free and liberated.

Now the first place, where we're caught right in the middle of the traffic and the confusion, is where many of us start our practice. That's where many of us see our relationships as being confusion, puzzlement, and bitterness, because we expect our relationship to be the *one* place that gives us peace from the traffic.

However, as we endeavor to practice with relationships, we begin to see that they are our best way to grow. In them we can see what our mind, our body, our senses, and our thoughts really are. Why are relationships such excellent practice? Why do they help us to go into what we might call the slow death of the ego?

Because, aside from our formal sitting, there is no way that is superior to relationships in helping us see where we're stuck and what we're holding on to. As long as our buttons are pushed, we have a great chance to learn and grow. So a relationship is a great gift, not because it makes us happy—it often doesn't—but because any intimate relationship, if we view it as practice, is the clearest mirror we can find.

You might say that relationships are the open door to our true self, to no-self. In our fear we always keep knocking at a painted door, one made of our dreams, our hopes, our ambitions; and we avoid the pain of the gateless gate, the open door of being with what is, whatever it is, here and now.

It's interesting to me that people don't see any connection between their misery and their complaints—their feeling of being a victim; the feeling that everyone is doing something *to* them. It's amazing. How many times has this connection been pointed out in dharma talks? How many? And yet because of our fear we won't look.

Only people of intelligence, energy, and patience will find that still point on which the universe turns. And unfortunately life for those who cannot or will not face this present moment is often violent and punishing; it's not nice; it doesn't care. Still, the truth is that it's not life, it's ourselves who are creating this misery. But if we really refuse to look at what we are doing—and I'm sorry how few people will look—then we're going to be punished by our life. And then we wonder why it's so hard on us. However, for those who patiently practice—sitting, sitting, sitting; who begin to practice steadily in their daily life—for those people there will be more and more a taste of the joy in a relationship in which no-self meets no-self. In other words openness meets openness. It's very rare, but it does happen. And when it happens I don't even know if we can use the word "relationship." Who is there to relate to whom? You can't say no-self relates to no-self. So for this state there are no words. And in this timeless love and compassion there is, as the Third Patriarch said, "No yesterday, no tomorrow, and no today."

Experiencing and Behavior

By experiencing, I mean that first moment when we receive life before the mind arises. For example: before I think, "Oh, that's a red shirt," there's just seeing. We could also speak of just hearing, just touching, just tasting, just thinking. This is the absolute: call it God, Buddha-nature, whatever you wish. This experience, filtered through my particular human mechanism, makes my world. We cannot point to anything in the world, seemingly inside or outside ourselves, which is not experiencing. But we couldn't have what we call a human life unless that experiencing were transformed into behavior. By behavior, I just mean the way something does itself. For example, as a human being, you do yourself: you sit, you move, you eat, you talk. In this sense even the rug has a behavior: the rug's behavior is just to lie there. (If we looked at it under a powerful microscope, we'd see it's by no means inert. It's a flood of energy, moving with incredible speed.)

So we can distinguish the arising—which is God, Buddha-nature, the absolute, just what is—from the world which is formed instantaneously, the other side of the arising. In fact, the two sides are one: the arising and what we call the world are not different. If we could really get this, we'd never again have any trouble in our lives, because it would be obvious that there is no past or future—and we'd see that all the stuff we worry about is nonsense.

For the most part we are only dimly aware of our experiencing. But we vaguely know that in some way our behavior and our experience are connected. If we have a headache and act irritably we probably realize that there is some connection between the pounding in our head and our irritable behavior. So even though we're not fully aware of our own experience, at least we do not view ourselves as divorced from our experience. But if other people are irritable, we may divorce their behavior from their experiencing. We can't feel their experience; and so we judge their behavior. If we think, "She shouldn't be so arrogant," we only see her behavior and judge it, because we have no awareness of what is true for her (her experiencing, her bodily sensations of fear). We slip into personal opinions about her arrogance.

Behavior is what we observe. We cannot observe experience. By the time that we have an observation about an event, it's past—and experience is never in the past. That's why the sutras say that we can't touch it, we can't see it, we can't hear it, we can't think about it—because the minute we attempt to do that, time and separation (our phenomenal world) have been created. When I observe my arm lifting, it's not me. When I observe my thoughts, they're not me. When I think, "This is me," I try to protect the "me." In fact, however, whatever I observe about myself (even though it's an interesting phenomenon with which I am closely associated) is not me. That's my behavior, the phenomenal world. Who I am is simply experiencing itself, forever unknown. The moment I name it, it is gone.

Behavior and experience are, however, not fundamentally separate. When I experience you (see you, touch you, hear you), you are my experiencing, just what is. But the human tendency is not to stop there; instead of you just being my experiencing, I add on to it my opinions about what you seem to be doing—and then I have separated myself from you. When the world seems separate, I think it has to be examined, analyzed, judged. When we live like this, rather than from experiencing itself, we are in trouble. We have to have memory, we have to have concepts; but if we don't understand their nature, if we don't use them properly, we create mayhem.

Like ourselves, other people are simply experiencing which looks like behavior. Yet we view them as behavior; we only see their behavior and are unaware of their experiencing. In truth experiencing is universal because that is what we are. When we can see the foolishness of our bondage to our thoughts and opinions, and increase the amount of time we live as experiencing, we are more able to sense the true life—the true experiencing—of another person. When we live a life that is not dominated by personal opinion but is instead pure experiencing, then we begin to take care of everyone, ourselves *and* others. Then we cannot view other people as objects, behavioral monkeys who are no more than their behavior.

All of practice is to return ourselves to pure experiencing. Out of that will emerge very adequate thinking and action. Usually we are unable to do this, however; instead we act in obedience to the thoughts and opinions that spin in our heads—which is backwards.

Nearly always we view other people as just behavior. We're not interested in the fact that their behavior cannot be divorced from their experiencing. With ourselves we get it to some degree, but not totally. In zazen, we see that only a fraction of ourselves is known to ourselves; and as that capacity for experiencing increases, our actions transform: they come not so much from our conditioning, our memories, but from life as it is, this very second.

This is true compassion. As we live more and more as our experiencing, we see that while we have a body and a mind that behave in certain ways, there is something (no-thing) in which the body and mind are held. We intuit that everyone is held in that way. Even though the behavior of another person may be irresponsible, and while we may have to oppose that behavior firmly, yet we and he or she are intrinsically the same. Only to the degree that we live a life of experiencing can we possibly understand the life of another. Compassion is not an idea or an ideal, it is a formless but all-powerful space that grows in zazen.

This space is always present. It's not something we have to hunt for and try to achieve. It is always what we are because it is our experiencing. We can't be other than this—but we can cover it with our ignorance. We don't have to "find" anything—which is why the Buddha said that after forty years he had achieved nothing. What is there to achieve? It's always right here.

Relationships Don't Work

I recently returned from Australia. I went there hoping to enjoy some normal weather—so the first two days it poured, which was fun. Then for the last five days of sesshin in Brisbane there was a cold gale. It was so strong that we could hardly stand up as we ran between buildings. We had to fight just to keep our balance. The wind was like a truck, roaring over the roof the whole time. Anyway, it was a good sesshin, and what I got (as I always do) is that no matter where you go, people are people: they are all wonderful and they are all troubled, as people are everywhere; and the same questions plague Australians as plague us. They have just as much difficulty with relationships as we do. So I want to talk for a few minutes about the illusions we have that relationships are going to work. See, they don't. They simply don't work. There never was a relationship that worked. You may say, "Well, why are we doing all this practice if that's true?" It's the fact that we want something to work that makes our relationships so unsatisfactory.

In a way life *can* work—but not coming from the standpoint that we are going to do something that will make it work. In everything we do in relation to other people, there is a subtle or not-so-subtle expectation. We think, "Somehow I'm going to figure this relationship out and make it work, and then I will get what I want." We all want something from the people we are in relationship to. None of us can say we don't want something from those we are in relationship to. And even if we avoid relationships that's another way of wanting something. So relationships just don't work.

Well, what does work then? The only thing that works (if we really practice) is a desire not to have something for myself but to support all life, including individual relationships. Now you may say, "Well, that sounds nice, I'll do that!" But nobody really wants to do that. We don't want to support others. To truly support somebody means that you give them everything and expect nothing. You might give them your time, your work, your money, anything. "If you need it, I'll give it to you." Love expects nothing. Instead of that we have these games: "I am going to communicate so our relationship will be better," which really means, "I'm going to communicate so you'll see what I want." The underlying expectation we bring to those games insures that relationships won't work. If we really see that, then a few of us will begin to understand the next step, of seeing another way of being. We may get a glimpse of it now and then: "Yes, I can do this for you, I can support your life and I expect nothing. Nothing."

There is a true story of a wife whose husband had been in Japan during the war. In Japan he lived with a Japanese woman and had a couple of children with her. He loved the Japanese woman very much. When he came home he did not tell his wife about this love. But finally, when he knew he was dying, he confessed to her the truth of the relationship and the children. At first she was very upset. But then something within her began to stir, and she worked and worked with her anguished feelings; finally, before her husband died, she said, "I will take care of them." So she went to Japan, found the young woman, and brought her and the two children back to the United States. They made a home together and the wife did all she could to teach the young woman English, to get her a job, and to help with the children. That's what love is.

A meditative practice is not some "airy-fairy" process, but a way of getting in touch with our own life. As we practice, more and more we have some idea of this other way of being, and we begin to turn away from a self-centered orientation—not to an "other-centered" orientation (because it includes ourselves), but to a totally open orientation. If our practice is not moving in that direction then it is not true practice. Whenever we want *anything* we know our practice has to continue. And since none of us can say other than that, it just means that for all of us our practice continues. I have been practicing a long time, yet I noticed that on this trip I just took (which was a long trip at my age, even though the sesshin was good, with strong impact on a lot of people) I was saying, "Well, it took too much out of me, I don't know if I will do it next year. Maybe I need more rest." The human mind is like that. Like anyone else, I want to be comfortable. I like to feel good. I don't like to be tired. And you may say, "Well, what's wrong with wanting a little comfort for yourself?" There is nothing wrong with wanting it, unless it is at variance with that which is more important to me than comfort, my primary orientation in life. If that primary orientation doesn't emerge from practice, then practice isn't practice. If we know our primary orientation, it will have its effect on every phase of life, on our relationships, our work, everything. If something doesn't emerge from practice that is more than just what *I want*, what would make my life more pleasant, then it's not practice.

But let's not oversimplify the problem. As we sit like this, we have to develop two, three, four aspects of practice. Just to sit in strong concentration has value. But unless we are careful, we can use it to escape from life. In fact, one can use the kind of power it develops in very poor ways. Concentration is one aspect of practice; we don't emphasize it here, but that ability must be acquired at some point. The Vipassana-type practice (which I prefer) in which you notice, notice, notice, is very valuable and I think the best and most basic training. Yet it can lead to people who are (as I think I was at one time) almost totally impersonal. There was nothing I felt emotionally because I had become an observing machine. That can sometimes be a drawback of this kind of practice. There are also other ways of practicing. Each way has strengths and each way has drawbacks. And there are various psychological and therapeutic trainings which are valuable; yet they also have drawbacks. The development of a human being into what I would call a balanced, wise, compassionate person is not simple.

In a relationship, whenever we sense unease—that point where it doesn't suit us—a big question mark should shoot up as to what is going on with us. How we can practice with the unease? I am not saying that all relationships should be continued forever, because the point of a relationship has nothing to do with the relationship itself. The point of relationship is the added power that life gets in working with it as a channel. A good relationship gives life more power. If two people are strong together, then life has a more powerful channel than it has with two single people. It's almost as though a third and larger channel has been formed. That is what life is looking for. It doesn't care about whether you are "happy" in your relationship. What it is looking for is a channel, and it wants that channel to be powerful. If it's not powerful life would just as soon discard it. Life doesn't care about your relationship. It is looking for channels for its power so it can function maximally. That functioning is what you are all about; all this drama about you and him or her is of no interest to life. Life is looking for a channel and, like a strong wind, it will beat on a relationship to test it. If the relationship can't take the testing, then either the relationship needs to grow in strength so that it can take it, or it may need to dissolve so something new and fresh can emerge from the ruins. Whether it crashes or not is less important than what is learned. Many people marry, for instance, when nothing is being served by their relationship. I am not advocating that people dissolve their marriages, of course. I simply mean that we often misinterpret what marriage is about. When a relationship isn't working, it means that the partners are preoccupied with "I": "What I want is . . ." or "This isn't right for me." If there is little wanting, then the relationship is strong and it will function. That's all life is interested in. As a separate ego with your separate desires, you are of no importance to life. And all weak relationships reflect the fact that somebody wants something for himself or herself.

These are big questions I am raising, and you may not agree with everything I'm saying. Still, Zen practice is about being selfless, about realizing that one is no-self. That does not mean to be a nonentity. It means to be very strong. But to be strong does not mean to be rigid. I've heard about a way of designing houses at the beach, where big storms can flood houses: when they are flooded, the middle of the house collapses and the water, instead of taking down the whole house, just rushes through the middle and leaves the house standing. A good relationship is something like that. It has a flexible structure and a way of absorbing shocks and stresses so that it can keep its integrity, and continue to function. But when a relationship is mostly "I want," the structure will be rigid. When it is rigid, it can't take pressure from life and so it can't serve life very well. Life likes people to be flexible so it can use them for what it seeks to accomplish.

If we understand zazen and our practice we can begin to get acquainted with ourselves, and how our troublesome emotions wreak havoc with our lives. If we really practice then very slowly, over the years, strength develops. At times this is a horrendous process. If anyone tells you differently they are not telling you about real meditation. Real meditation is by no means a flowery, blissful process. But if we really do it in time we begin to know what it is we're after; we begin to see who we are. So I want you to appreciate your practice and really do it. Practice is not a trimming on your life. Practice is the foundation. If that's not there nothing else will be there. So let's keep clarifying what our practice is at this moment. And who knows, some of us might even find ourselves in a relationship that works—one that has a very, very different base. It is up to us to create that base. So let's just do that.

Relationship Is Not to Each Other

We sit sesshin in order to know who we are. We have a mind and body; but those elements don't explain the life we are. Shakespeare's Polonius said, "To thine own self be true, and it must follow, as the night the day, thou canst not then be false to any man." We want to know our *true self*. We may form a picture of something called "the true self," as if it were an actual entity, floating about. We are in sesshin to discover, to *be* our true self. But what on earth is it?

If you had to define "true self" what would you say? Let's think for a moment. What am I coming up with? Something like: the functioning of a man or a woman in which there is no self-centered motivation. It's not hard to see that such a person would not be human in any familiar sense. From a different point of view such a person would be fully human—but not in the way we usually think of ourselves and others. Such a person would actually be nobody at all.

As we struggle though life and sense the shortcomings of our relationships to this person or to that person, or to our work or a particular activity, one of our blinding errors is the idea that "I am related to that person or event." For instance, suppose I'm married. The usual way we think of marriage is, "I am married to him." But as long as I say, "I am married *to* him," there are two of us and in true self there cannot be two. True self knows no separation. It may *look* like me being married to him, but true self—call it the infinite energy potential—knows no separation. True self forms into different shapes but essentially it remains one self, one energy potential. When I say I am married to you, or I own a Toyota, or I have four children, in everyday language that is so. But we need to see that it's not the real truth. In fact, I am not married *to* somebody or something; I *am* that person, I *am* that thing. The true self knows no separation.

Now you may say that's all very pretty; but practically speaking, what do we do about the difficult problems that occur in our lives? We all know that work can present enormous challenges; so can children, parents, any relationship. Suppose I'm married to someone who is extremely difficult—not just a little difficult, but extremely difficult. Suppose the children of the marriage are suffering. I've often talked about the fact that when we are suffering, we must become the suffering. That's how we grow, true enough. But now does that apply when a situation is so difficult that everyone involved is taking a beating? What do we do? And there are many variations on relationship problems. Suppose that I have a partner who is deeply committed to a field of study, and the only place the study can be carried out is in Africa for three or four years—and my work keeps me here. Then what? What do I do? Or perhaps I have aged parents who need my care—yet my profession, my responsibilities and obligations, call me elsewhere, what do I do? Such problems are what life is made of. Not all problems are as tough as these, but less demanding ones may still send us up the wall with worry.

In any situation our devotion should be not to the other person *per se*, but to the true self. Of course the other person embodies the true self, yet there is a distinction. If we are involved in a group, our relationship is not to the group, but to the true self of the group. By the "true self" I'm not talking about some mystical ghost that floats above. True self is nothing at all; and yet it's the only thing that should dominate our life; it is the only Master. Doing zazen, or sitting sesshin, is for the purpose of better understanding our true self. If we don't understand it, then we will be confused forever by problems and won't know what to do. The only thing to be served is not a teacher, not a center, not a job, not a mate, not a child, but our true self. So how do we know how to do that? It's not easy and it takes time and perseverance to learn.

Practice makes it obvious that in almost all of our life we are not greatly interested in our true self; we're interested in our *small* self: we are interested in what *we* want, what *we* think, what *we* hope for, what would make us feel good, what would ensure our health, our well-being; that's where our energy goes. An intelligent practice slowly illuminates that fact. And it's not good or bad that we *are* like that, it's just the way it is. When we have some illumination of our usual self-centered activity, when we are aware of the grief and the agony that it produces, sometimes we can turn away from it. We may even get a glimpse of another way of being: the true self.

In a concrete situation, what is the way to serve the true self? The way may look very rough, very unkind—and sometimes it will be the opposite. There are no formulas. Perhaps I give up my good job in New York and stay at home to take care of my parents. Or perhaps I don't. No one but my true self can tell me what I should do. If our practice has matured to a point where we don't often fool ourselves — because we are in touch with our actual experience — then we know more and more what is the compassionate action to take. When we are nobody, no-self (and we'll never be that completely) the right action is obvious.

All relationships can teach us something; and some of them, sadly, must come to an end. There may come a time when the best way to serve the true self is to move on. No one can tell me what is best; no one knows except my true self. It doesn't matter what my mother says about it, or what my aunt says about it; in a certain sense it doesn't even matter what I say about it. As one teacher says, "Your life is none of your business." But our practice is definitely our business. And that practice is to learn what it means to serve that which we cannot see, touch, taste or smell. Essentially the true self is no-thing, and yet it is our Master. And when I say it's no-thing, I don't mean nothing in the ordinary sense; the Master is not a thing, yet it's the only thing. When we're married, we're not married to each other, but to the true self. When we teach a group of children, we're not teaching the children, we are expressing the true self in a way appropriate to the classroom.

Now this may sound idealistic and remote; yet every five minutes we get a chance to work with it. For example: the interchange with someone who irritates us; the little encounter that goes sour, when we feel they should "know better"; the irritation when my daughter says she'll telephone — and she forgets. What is the true self in all of these incidents? Usually we can't see the true self; we can only see how we miss it. We can be aware of irritability, annoyance, impatience. And such thoughts we can label. We can patiently do that, we can experience the tension the thoughts generate. In other words we can experience what we put *between* ourselves and our true self. When such careful practice is put first in our life we serve the Master — and then we grow in knowledge of what must be done.

There is only one Master. The Master is not me, nor anyone else, not Sabba Somebody or Guru Somebody; no person can be the Master. And no Center is anything but a tool of the Master. No marriage, no relationship, is anything but that. But to realize that fact we have to illuminate our activity not once, but ten thousand times. We have to put a searchlight on our unkind thoughts about people and situations. We must make conscious how we feel, what we want, what we expect, how terrible we think someone else is, or how terrible we are—that cloud over everything. We're like a little squid that produces a flood of ink so our mischief can't be seen. When we wake up in the morning we immediately start squeezing out ink. What is our ink? Our self-centered preoccupation, which clouds the water around us. When we live self-centered lives we create trouble. We may insist we don't like horrible fairy tales, but we *do* like them. Something within us is fascinated by our drama and so we cling to it and confuse ourselves.

True practice brings us more and more into that plain and undramatic space in which things are just as they are—just functioning. And that functioning cannot come from self-centeredness. Sitting in sesshin greatly increases our chances of spending more of our life in that plain space. But we have to have patience, persistence, and posture; we must maintain equanimity and *sit*. True self is nothing at all. It's the absence of something else. An absence of what?

V. SUFFERING

True Suffering and False Suffering

Yesterday I was talking to a friend who recently had a major operation and has been recovering. I asked her what would be a good subject for a dharma talk, and she laughed and said, "Patience and Pain." She found it very interesting that in the days immediately following her operation the pain was clear, clean, and sharp; and it was no problem. Then, as she became a little stronger, the mind began to work—and the suffering began. All her thoughts *about* what was happening to her began to appear.

In a way we sit for no purpose; that's one side of practice. But the other side is that we want to be free from suffering. Not only that, but we want others to be free from suffering. So a key point in our practice is to understand what suffering is. If we really understand suffering we see how to practice, not just while sitting, but in the rest of our life. We can understand our daily life and see that it's really not a problem. A few weeks ago someone gave me an interesting article on suffering, and the first part of it was about the meaning of the word—"suffering." I'm interested in these meanings; they are teachings in themselves.

The writer of this article pointed out that the word "suffering" is used to express many things. The second part, "fer," is from the Latin verb *ferre* meaning "to bear." And the first part, "suf," is from *sub*, meaning "under." So there's a feeling in the word "to be under," "to bear under," "to totally be under"—"to be supporting something from underneath."

Now, in contrast, the words "affliction," "grief," and "depression" all bring images of weight; of something bearing *down upon* us. In fact the word "grief" is again from the Latin *gravare*, which means "press down."

So there are two kinds of suffering. One is when we feel we're being pressed down; as though suffering is coming at us from without, as though we're receiving something that's making us suffer. The other kind of suffering is being under, just bearing it, just *being* it. And this distinction in understanding suffering is one of the keys to understanding our practice.

I've sometimes distinguished between "suffering" and "pain," but now I'd like to use just the word "suffering" and distinguish between what I call *false* suffering and *true* suffering. That difference in understanding is very important. The foundation of our practice, and the first of the Four Noble Truths, is the statement of the Buddha that "Life is suffering." He didn't say it's suffering sometimes—he said life *is* suffering. And I want to distinguish between those two kinds of suffering.

Often people will say, "I certainly can see that life is suffering when everything is going wrong, and everything's unpleasant, but I really don't get it when life is going along well and I'm feeling good." But there are different categories of suffering. For instance, when we don't get something we want, we suffer. And yet when we *do* get it we also suffer, because we know that if we get it we can lose it. It doesn't matter whether we get it or don't get it, if it happens to us or doesn't happen to us. We suffer because life is constantly changing. We know we can't hold on to the pleasant things, and we know that even if unpleasantness disappears, it can come again.

The word "suffer" doesn't necessarily imply a dramatic major experience; even the nicest day is not free of suffering. For instance, you might have the best breakfast you can imagine, you might see just the friend you want to see, you might go to work and have everything go smoothly. There aren't many days as nice as that; but even so, we know that on the next day it could be just the opposite. Life presents us with no guarantees; and because we know that, we're uneasy and anxious. If we truly examine our situation from the usual point of view, life is suffering, like an affliction.

Now my friend noticed that when there was just the physical pain, there was no problem. The minute she began to have thoughts about the pain, she began to suffer and be miserable. It makes me think of a few lines of Master Huang Po: "This mind is no mind of conceptual thought and it is completely detached from form. So Buddhas and sentient beings do not differ at all. If you can only rid yourselves of conceptual thought, you will have accomplished everything. But if you students of the Way do not rid yourselves of conceptual thought in a flash, even though you strive for aeon and aeon, you will never accomplish it."

It's the play of our minds, of conceptualization about anything that happens to us, that is the problem. There's nothing wrong with conceptualization *per se*; but when we take our opinions about any event to be some kind of absolute truth and *fail to see* that they are *opinions*, then we suffer. That's false suffering. "A tenth of an inch of difference, and heaven and earth are set apart."

Now a point to add here is that it doesn't make any difference what's happening: It may be very unfair; it may be very cruel. All of us have things happen to us that are unfair, mean, and cruel. And our usual way is to think, "This is terrible!" We fight back, we oppose the event. We try to do as Shakespeare said: "To take arms against a sea of troubles, and by opposing, end them."

It would be nice if it *did* end "the slings and arrows of most outrageous fortune." Day by day we all meet events that seem to be most unfair, and we feel that the only way to handle an attack is to fight back; and the way we fight is with our minds. We arm ourselves with our anger and our opinions, our self-righteousness, as though we were putting on a bulletproof vest. And we think this is the way to live our life. All that we accomplish is to increase the separation, to escalate the anger, and to make ourselves and everyone else miserable. So, if this approach doesn't work, how *do* we handle the suffering of life? There's a Sufi story about this.

At one time there was a young man whose father was one of the greatest teachers of his generation, respected and revered by everyone. And this young man, having grown up hearing his father speak great words of wisdom, felt that he knew all there was to know. But his father said, "No, I can't teach you what you need to know. The person I want you to go and see is a peasant teacher, a man who is illiterate, just a farmer." The young man wasn't pleased with this, but he went off and traveled on foot, not very willingly, until he came to the village where this peasant lived. It happened at this time that the peasant was on his way by horseback from his own farm to another farm, and he saw the young man coming toward him.

When the young man came near and bowed before him, the teacher looked down and said, "Not enough."

Thereupon the young man bowed to his knees, and the peasant teacher again said, "Not enough." Then he bowed to the horse's knees, but again the teacher said, "Not enough." So the young man bowed once more, this time to the horse's feet, touching the horse's hoof. Then the peasant teacher said to him, "You can go back now. You have had your training." And that was all.

So (remembering the definition of the word "suffer") until we bow down and bear the suffering of life—not opposing it, but *absorbing* it and *being* it—we cannot see what our life is. This by no means implies passivity or nonaction, but action from a state of complete acceptance. Even "acceptance" is not quite accurate—it's simply *being* the suffering. It isn't a matter of protecting ourselves, or accepting something else. Complete openness, complete vulnerability to life, is (surprisingly enough) the only satisfactory way of living our life.

Of course, if you're anything like me, you'll avoid it as long as possible—because it's one thing to talk about, but extremely difficult to do. Yet when we do it, we know in our very guts who we are and who everyone else is; and the barrier between ourselves and others is gone.

Our practice throughout our lifetime is just this: At any given time we have a rigid viewpoint or stance about life; it includes some things, it excludes others. We may stick with it for a long time, but if we're sincerely praticing our practice itself will shake up that viewpoint; we can't maintain it. As we begin to question our viewpoint we may feel struggle, upset, as we try to come to terms with this new insight into our life; and for a long time we may deny it and struggle against it. That's part of practice. Finally we become willing to experience our suffering instead of fighting it. When we do so our standpoint, our vision of life, abruptly shifts. Then once again, with our new viewpoint, we go along for a while—until the cycle begins anew.

Once again the unease comes up, and we have to struggle, to go through it again. Each time we do this—each time we go into the suffering and let it be—our vision of life enlarges. It's like climbing a mountain. At each point that we ascend we see more; and that vision doesn't deny anything that's below—it includes that—but it becomes broader with each cycle of climbing, of struggle. And the more we see, the more expansive our vision, the more we know what to do, what action to take.

In talking to many, many people, the main thing I notice is that they don't understand suffering. Of course I don't always either; I try to avoid it as hard as anyone. But to have some theoretical understanding of what suffering is and how to practice with it is immensely useful, especially in sesshin. We can better understand what sesshin is and how best to use it and really practice.

This mind that creates false suffering is operating constantly in sesshin. There isn't one of us who isn't subject to it. I noticed it myself last night. I could hear my mind complaining: "What, another sesshin?! You just did a sesshin last weekend!" Our minds work that way. Then, seeing that nonsense, we remind ourselves, "What do I really want for myself or anyone else?" Then our mind quiets down again.

So as we do zazen we patiently refuse the domination of these thoughts and opinions about ourselves, about events, about people; and we constantly turn back to the only certain reality: this present moment. Doing that, our focus and samadhi constantly deepen. So in zazen the bodhisattva's renunciation is that practice, that turning away from our fantasy and our personal dream into the reality of the present. And in sesshin, each moment that we practice like that gives us what we can't get in any other way, a direct knowledge of ourself. Then we are facing this moment directly; we're facing the suffering. And when we're really, finally, willing to settle into it, just be it, then we know, and no one needs to tell us, what we are, and what everything else is.

Now, sometimes people say, "It's too hard." But in fact, not practicing at all is much, much harder. We really fool ourselves when we don't practice. So please be very clear with yourself about what must be done to end suffering; and also that by practicing with such courage we can enable others to have no fear, no suffering. We do it by the most intelligent, patient, persistent practice. We *never* do it by our complaints, our bitterness and anger; and I don't mean to suppress them. If they come up, notice them; you don't have to suppress them. Then *immediately* go back into your breath, your body, into just sitting. And when we do that there is not one of us who, by the end of the sesshin, will not find the rewards that real sitting gives. Let's sit like that.

Renunciation

Suzuki Roshi said, "Renunciation is not giving up the things of this world, but accepting that they go away." Everything is impermanent; sooner or later everything goes away. Renunciation is a state of nonattachment, acceptance of this going away. Impermanence is, in fact, just another name for perfection. Leaves fall; debris and garbage accumulate; out of the debris come flowers, greenery, things that we think are lovely. Destruction is necessary. A good forest fire is necessary. The way we interfere with forest fires may not be a good thing. Without destruction, there could be no new life; and the wonder of life, the constant change, could not be. We must live and die. And this process is perfection itself.

All this change is not, however, what we had in mind. Our drive is not to appreciate the perfection of the universe. Our personal drive is to find a way to endure in our unchanging glory forever. That may seem ridiculous, yet that's what we're doing. And that resistance to change is not attuned with the perfection of life, which is its impermanence. If life were not impermanent, it couldn't be the wonder that it is. Still, the last thing we like is our own impermanence. Who hasn't noticed the first gray hair and thought, "Uh-oh." So a battle rages in human existence. We refuse to see the truth that's all around us. We don't really see life at all. Our attention is elsewhere. We are engaged in an unending battle with our fears about ourselves and our existence. If we want to see life we must be attentive to it. But we're not interested in doing that; we're only interested in the battle to preserve ourselves forever. And of course it's an anxious and futile battle, a battle that can't be won. The one who always wins is death, the "right-hand man" of impermanence.

What we want out of life as we live it is that others reflect our glory. We want our partners to ensure our security, to make us feel wonderful, to give us what we want, so that our anxiety can be eased for a little while. We look for friends who will at least take the cutting edge off of our fear, the fear that we're not going to be around one day. We don't want to look at that. The funny thing is that our friends are not fooled by us; they see exactly what we're doing. Why do they see it so clearly? Because they're doing it too. They're not interested in *our* efforts to be the center of the universe. Yet we wage the battle ceaselessly. We are frantically busy. When our personal attempts to win the battle fail, we may try to find peace in a false form of religion. And people who offer that carrot get rich. We are desperate for anyone who will tell us, "It's all right. Everything can be wonderful for you." Even in Zen practice we try to find a way around what practice really is, so that we can gain a personal victory.

People often say to me, "Joko, why do you make practice so hard? Why don't you hold out any cookies at all?" But from the point of view of the small self, practice can only be hard. Practice annihilates the small self, and the small self isn't interested in that one bit. It can't be expected to greet this annihilation with joy. So there's no cookie that can be held out for the small self, unless we want to be dishonest.

There is another side to practice, however: As our small self dies—our angry, demanding, complaining, maneuvering, manipulative self—a real cookie appears: joy and genuine self-confidence. We begin to taste what it feels like to care about someone else without expecting anything in return. And this is true compassion. How much we have it depends on the rate at which the small self dies. As it dies, here and there we have moments when we see what life is. Sometimes we can spontaneously act and serve others. And with this growth always comes repentance. When we realize that we have almost constantly hurt ourselves and others, we repent—and this repentance is itself pure joy.

So let us notice that our efforts in sesshin are to perfect ourselves: we want to be enlightened, we want to be clear, we want to be calm, we want to be wise. As our sitting settles down into the present moment we say, "Isn't this boring! —the cars going by, feel my knees hurting, notice my tummy growling . . ." We have no interest in the infinite perfection of the universe. In fact the infinite perfection of the universe might be the person sitting next to us who breathes noisily or is sweaty. The infinite perfection is this being inconvenienced: "I'm not having it my way at all." At any moment there is just what's happening. Yet we're not interested in that. Instead we're bored. Our attention goes in another direction. "Forget reality! I'm here to be enlightened!"

But Zen is a subtle practice: even as we fight it and resist it and distort it, our concepts about it tend to destroy themselves. And slowly, in spite of ourselves, we begin to be interested in what practice really is, as opposed to our ideas of what we think it should be. The point of practice is exactly this clashing space in which my desires for my personal immortality, my own glorification, my own control of the universe, clash with what is. This moment occurs frequently in our lives: the moment when we feel irritability, jealousy, excitement—the clash between the way I want it and the way it is. "I hate her noisy breathing. How can I be aware of *what is* when she does that?" "How can I practice when the boys next door play rock and roll?" Every moment offers us a wealth of opportunities. Even on the calmest, most uneventful day we get many opportunities to see the clash between what we want and the way it really is.

All good practice aims to make our false dreams conscious, so that there is nothing in our physical and mental experience that is unknown to us. We need not only to know our anger, we need to know our own personal ways of handling our anger. If a reaction is not conscious we can't look at it and turn away from it. Each defensive reaction (and we have one about every five minutes) is practice. If we practice with the thoughts and physical sensations that comprise that reaction, we open to wholeness, or holiness, if you want to call it that. In good practice we are always transforming from being personally centered (caught in our personal reactions) to being more and more a channel for universal energy, this energy that shifts the universe a million times a second. In our phenomenal lives what we see is impermanence; the other side is something else; we won't give it a name. When we practice well we are increasingly a channel for this universal energy, and death loses its sting.

A major obstacle to seeing is unawareness that all practice has a strong element of resistance. It is bound to have this unwillingness until our personal self is completely dead. Only a Buddha has no resistance, and I doubt that in the human population there are any Buddhas. Until we die we always have some personal resistance that has to be acknowledged.

A second major obstacle is a lack of honesty about who we are at each moment. It's very hard to admit, "I'm being vengeful" or "I'm being punishing" or "I'm being self-righteous." That kind of honesty is hard. We don't always have to share with others what we observe about ourselves; but there should be nothing going on that we're not aware of. We have to see that we are chasing ideals of perfection rather than acknowledging our imperfection.

A third obstacle is being impressed and sidetracked by our little openings as they occur. They're just the fruit; they have no importance unless we use them in our lives.

A fourth obstacle is having little understanding of the magnitude of the task that we have embarked upon. The task is not impossible, it's not too difficult; but it is unending.

A fifth obstacle, common among people who spend much time

at Zen centers, is substituting talk and discussion and reading for persistent practice itself. The less we say about practice, the better. Outside of a direct student-teacher setting, the last thing that I will talk about is Zen practice. And I don't talk about the dharma. Why talk about it? My job is to notice how I violate it. You know the old saying, "He who knows does not say, and he who says does not know." When we talk about practice all the time, our talk is another form of resistance, a barrier, a cover. It's like academics who save the world every night at the dinner table. They talk and talk and talk—but what difference does it make? At the other end of that pole would be someone like Mother Teresa. I don't think she does much talking. She is busy *doing*.

Intelligent practice always deals with just one thing: the fear at the base of human existence, the fear that *I am not*. And of course I am not, but the last thing I want to know is that. I am impermanence itself in a rapidly changing human form that appears solid. I fear to see what I am: an ever-changing energy field. I don't want to be that. So good practice is about fear. Fear takes the form of constantly thinking, speculating, analyzing, fantasizing. With all that activity we create a cloudy cover to keep ourselves safe in a make-believe practice. True practice is *not* safe; it's anything but safe. But we don't like that, so we obsess with our feverish efforts to achieve our version of the personal dream. Such obsessive practice is itself just another cloud between ourselves and reality. The only thing that matters is seeing with an impersonal search-light: seeing things are they are. When the personal barrier drops away, why do we have to call it anything? We just live our lives. And when we die, we just die. No problem anywhere.

It's OK

Enlightenment is the core of all religion. But we have quite often a strange picture of what it is. We equate the enlightened state with a state in which we have become quite perfect, quite nice and quiet, calm and accepting. And that's not it.

I'm going to ask a series of questions about certain unpleasant states. I am not saying not to try to prevent these states, not to change them; I'm not saying we should not have strong feelings or preferences about them. Nevertheless from these examples we can begin to get a clue; and when we have a clue we can see more clearly what we're doing in practice. Here are the questions:

- If I am told "Joko, you have one more day to live," is that OK with me? Or if someone told you that, is it OK with you?
- If I am in a severe accident, and my legs and arms have to be amputated, is this OK with me? If that were to happen to you, is it OK?
- If I were never again to receive a kind or friendly encouraging word from anyone, is this OK with me?
- If I, for whatever reason, have to be bedridden and in pain for the rest of my life, is this OK with me?
- If I make a complete fool of myself, in the worst possible circumstances, is this OK with me?
- If the close relationship that you dream of and hope for never comes to pass, is this OK?
- If for whatever reason I have to live out my life as a beggar, with little food and no shelter, exposed to the cold, is this OK with me? With you?
- If I must lose whatever or whomever I care for, is it OK with me?

Now, I can't answer OK to any of those. And if you're honest, I don't think that any of you can either. But to answer "OK" is the enlightened state, if we understand what it means for something to be OK. For something to be OK, it doesn't mean that I don't scream, or cry, or protest, or hate it. Singing and dancing are the voice of the dharma, and screaming and moaning are also the voice of the dharma. For these things to be OK for me doesn't mean that I'm happy about them. If they're OK, what does that mean? What *is* the enlightened state? When there is no longer any separation between myself and the circumstances of my life, whatever they may be, that is it.

Of course, I presented a particularly unpleasant set of options. I might ask instead, "If you had to receive a billion dollars, is that OK with you?" And you might say, "Oh yes!" Yet to have a billion dollars presents almost as many difficulties as to be a beggar. In any case the question is whether it is OK with you to live with the circumstances that life brings to you. That doesn't mean blind acceptance. It doesn't mean if you're ill, not to do all you can to get well; but sometimes things are inevitable—there's very little we can do. Then is it OK?

You may protest that a person for whom any condition is OK is not human. And in a way you're right; such a person is not human. Or we might say they are truly human. We can say it either way. But a person who has no aversion to any circumstance is not a human being as we usually know human beings. I've known a few people who approximated this condition. And this is the enlightened state: the state of a person who, to a great degree, can embrace any or all conditions, good or bad. I'm not talking about a saint; I'm talking about that state (often preceded by enormous struggle) when it's OK. For instance, we often wonder how we will die. The key is not to learn to die bravely, but to learn not to *need* to die bravely. We may have that in small areas, but mostly we wish to be something other than we are. An interesting attitude indeed: not to learn to put up with any circumstance, but to learn not to *need* a particular attitude toward a circumstance.

Most therapies are intended to adjust my needs and wants to your needs and wants, to foster some peace between us. But suppose I have no objection to any of my needs or wants, or to any of your needs or wants—it's all perfect as it is—then what needs to be adjusted? You may say that a person who could answer any of the questions with "Yes" would be weird. I don't think so. If you met such a person you wouldn't notice anything odd. You would probably notice immense peace in being with that person. Someone who has little self-concern, who is willing to be as he or she is and everything else to be as it is, is truly loving. You would find such a person to be very supportive if that were appropriate, or quite nonsupportive if that were appropriate. And this person would know the distinction, would know what to do, because this person would be you.

So I want you to consider: what is the basis for you to be able to answer any condition of life with, "It's OK. I have no complaints whatsoever"? This doesn't mean that you are never upset, but there is a basis in which all of life rests, so that no matter what you can answer, "It's OK." And what we're doing in our practice (whether you know it or not or whether you want to or not) is learning to know this basis, this fact that can enable us to say in time: "It's OK." Or, as in the Lord's prayer: Thy will be done.

One way to evaluate our practice is to see whether life is more and more OK with us. And of course it's fine when we can't say that, but still it is our practice. When something's OK with us we accept everything we are with it; we accept our protest, our struggle, our confusion, the fact that we're not getting anywhere according to our view of things. And we are willing for all those things to continue: the struggle, the pain, the confusion. In a way that is the training of sesshin. As we sit through it an understanding slowly increases: "Yes, I'm going through this and I don't like it—wish I could run out—and somehow, it's OK." That increases. For example: you may enjoy life with your partner, and think, "Wow, this is the one for me!" Suddenly he or she leaves you; the sharp suffering and the experience of that suffering is the OKness. As we sit in zazen, we're digging our way into this koan, this paradox which supports our life. More and more we know that whatever happens, and however much we hate it, however much we have to struggle with it—in some way it's OK. Am I making practice sound difficult? But practice *is* difficult. And strangely enough, those who practice like this are the people who hugely enjoy life, like Zorba the Greek. Expecting nothing from life, they can enjoy it. When events happen that most people would call disastrous, they may struggle and fight and fuss, but still they enjoy—it's OK.

Unless we completely misunderstand what sesshin practice is, more and more we appreciate the struggles, the weariness and pain, even as we dislike them. And let's not forget the wonderful moments of sesshin: then our joy and appreciation may startle us. For such practice, a residue builds which is *understanding*. I'm not as interested in the enlightenment experiences as I am in the practice which builds this understanding, because as it grows, our life changes radically. It may not change in the way we expect it to change. We grow in understanding and appreciation of the perfection of each moment: our aching knees and back, the itch on our nose, our sweat. We grow in being able to say, "Yes, it's OK." The miracle of sitting zazen is this miracle of appreciation.

It would be very hard for me if I never again were to receive a kind or friendly word. Is this OK with me? Of course it's not, but what would the practice be? If I were kidnapped in some barbaric country and shut away in a jail, what would the practice be? Things that drastic don't happen to most of us. But on a lesser scale, disasters do happen and our pictures of how our lives should be are blown away. Then we have a choice: do we face the disaster directly and make it our practice or do we run once again, learning nothing and compounding our difficulties? If we want a life that's peaceful and productive, what do we need? We need the ability (which we learn slowly and unwillingly) to be the experience of our life as it is. Most of the time I don't want to do that, and I suspect that you don't either. But that is what we're here to learn. And surprisingly, we are learning it. Almost everyone after sesshin is happier. Maybe because it's over, but not just that. After sesshin just a walk down the street is great. It wasn't great before sesshin; but it's great after sesshin. This attitude may not last very long. Three days later we're already searching for the next solution. Still we *have* learned a little about the error of this kind of search. The more we have experienced life in all its guises as being OK, the less we are motivated to turn away from it in an illusory search for perfection.

Tragedy

According to the dictionary a tragedy is "a dramatic or literary work depicting a protagonist engaged in a morally significant struggle ending in ruin or profound disappointment." From the usual point of view life *is* a tragedy—yet we spend our lives in a hopeless attempt to hide from the tragedy. Each of us is a protagonist playing our leading role on our little stage. Each of us feels we are engaged in a morally significant struggle. And—though we don't want to admit this—that struggle will inevitably end in our ruin. Aside from any accident we might encounter in life, there's one "accident" at the end that none of us can avoid. We're done for. From the moment of conception our life is on its way out. And from a personal point of view this is a tragedy. So we spend our life in a pointless battle to avoid that end. That misdirected battle is the real tragedy.

Suppose we live near the ocean in a warm climate where we can swim all year, but in water where there are sharks. If we're smart swimmers we'll research where the sharks are likely to hang out, and then we'll avoid that area. But sharks being sharks, sooner or later one may stray into our swimming area and find us. We can never be certain. If a shark doesn't get us the riptide may. We may swim a lifetime and never encounter a shark; yet worry about them can poison all our swimming days.

Each one of us has figured out where we think the sharks are in our lives, and we spend most of our energy in worrying about them. It's sensible to take precautions against physical injury; we buy insurance, we have our children vaccinated, we try to lower our cholesterol level. Still an error creeps into our thinking. What is that error?

What is the difference between taking reasonable precautions, and ceaseless worry and mind-spinning? There's a famous Buddhist parable: a man was being chased by a tiger. In his desperation he dove over the side of a cliff and grabbed a vine. As the tiger was pawing away above him he looked below and saw another tiger at the base of the cliff, waiting for him to fall. To top it off two mice were gnawing away at the vine. At that moment he spotted a luscious strawberry and, holding the vine with one hand, he picked the strawberry and ate it. It was delicious! What finally happened to the man? We know, of course. Is what happened to him a tragedy?

Notice that the man being chased by a tiger didn't lie down and say, "Oh, you beautiful creature. We are one. Please eat me." The story is not about being foolish—even though on one level, the man and the tiger *are* one. The man did his best to protect himself, as we all should. Nevertheless, if we're left hanging on that vine, we can either waste that last moment of life or we can appreciate it. And isn't every moment the last moment? There is no moment other than this.

It is sensible to take care of our own mind and body; the problem arises with our *exclusive* identification with them. A few individuals in the history of humankind have been identified as much with other forms of life as with their own. For them there is no tragedy, because for them no adversary exists. If we are one with life—no matter who it is, or what it is, or what it does—there is no protagonist, no adversary, no tragedy. And the strawberry can be appreciated.

When our practice is steady, ongoing, intense, we can begin to sense the error of an exclusive identification with mind and body. (Of course we will see this to varying degrees and sometimes we will not see it at all.) It's not an intellectual comprehension. Modern physics makes it clear that we are "one," that we are just different manifestations of one energy, and that's not hard to comprehend intellectually. But as human beings with minds, bodies, and emotions, how much do we know that in every cell of our bodies?

When our identification with mind and body is loosened and to some degree seen for what it is, we become more open to the concerns of others, even when we don't agree with them, even when we have to oppose them. Increasingly our attitude can include the other side of the picture, the point of view of the other person.

When that happens there is no longer a protagonist opposing an adversary.

Practice is more and more to see through the fiction of this exclusive identification, the conceptual disease which rules our actions. As we do zazen we have a precious opportunity to face ourselves, to see the nature of the false thinking which creates the illusion of a separate self.

The immense cunning of the human mind can operate very well when not challenged; but under the assault of sesshin, sitting motionless for long hours, the dishonesty and evasions of the mind become crystal clear. And the tension created by the cunning mind also begins to be felt. It may take us aback to realize that nothing outside of ourselves is attacking us. We are only assaulted by *our* thoughts, *our* needs, *our* attachments, all born from our identification with our false thinking which in turn creates a closed-in, separate, miserable life. In daily sitting we can sometimes avoid this realization. But it's hard to avoid it when we sit eight hours a day; and the more days we sit, the harder it is.

As we patiently practice (experiencing our breath, being aware of the thinking process) realization is born, not intellectually, but in the very cells of our body. False thinking evaporates as clouds in the heat of the sun, and we find in the midst of our suffering an openness, a spaciousness and joy we have never known before.

Someone once insisted to me, "That still doesn't handle the problem of death. We still die." And we do. But if in the second before death we can say, "Oh, what a delicious strawberry!"—then there is no problem. If the shark eats us, then the shark has had a good meal. And perhaps in time a fisherman will catch the shark. From the shark's point of view it's a tragedy. From life's viewpoint, no.

I'm not suggesting a new ideal for us to chase after. The man running away from the tiger, shaking with fear, *is* the dharma. Whatever you are is the dharma. So as you sit, and as you struggle and feel miserable or confused, just be that. If you are blissful, just be that—but don't cling to it. Then each moment is just what each moment is. With such patient practice, we see the error of our exclusive identification with our mind and body; we begin to understand.

Tragedy always involves a   protagonist engaged in a struggle. But we don't have to be a protagonist, engaged in an endless struggle with forces external to ourselves. The struggle is our own interpretation, ending in ruin only if we see it as such. As the Heart Sutra says, "No old age and death and no end to old age and death. . . . No suffering and no end to suffering." The man being chased by the tiger is finally eaten. OK. No problem.

The Observing Self

> "Who is there?" asks God.
> "It is I."
> "Go away," God says . . .
> Later . . .
> "Who is there?" asks God.
> "It is Thou."
> "Enter," replies God.

What we ordinarily think of as the self has many aspects. There is the thinking self, the emotional self, and the functional self which does things. These together comprise our describable self. There is nothing in those areas that we cannot describe; for instance, we can describe our physical functioning: we take a walk, we come home and we sit down. As for emotion: we can usually describe how we feel; when we get excited or upset, we can say that our emotion arises, peaks, and falls in intensity. And we can describe our thinking. These aspects of the describable self are the primary factors of our life: our thinking self, our emotional self, our functional self.

There is, however, another aspect of ourself that we slowly get in touch with as we do zazen: *the observing self*. It is important in some Western therapies. In fact, when used well, it is why the therapies work. But these therapies do not always realize the radical difference between the observing self and other aspects of ourselves, nor do they understand its nature. All the describable parts of what we call ourselves are limited. They are also linear; they come and go within a framework of time. But the observing self cannot be put in that category, no matter how hard we try. That which observes cannot be found and cannot be described. If we look for it there is nothing there. Since there is nothing we can know about it, we can almost say it is another dimension.

In practice we observe — or make conscious — as much as we can of our describable selves. Most therapies do this to some degree; but zazen, continued for years, cultivates the observing self more deeply than do most therapies. As we practice we must observe how we work, how we make love, how we are at a party, how we are in a new situation with strange people. There is nothing about ourselves that shouldn't come under scrutiny. It's not that we stop other activities. Even when we are completely absorbed in our daily life the observing function continues. Any aspect of ourselves that is *not* observed will remain muddy, confusing, mysterious. It will seem independent of us, as though it is happening all by itself. And then we will get caught in it and carried away into confusion.

At one time or another all of us get carried away by some kind of anger. (By "anger" I mean also irritability, jealousy, annoyance, even depression.) In years of sitting we slowly uncover the anatomy of anger and other emotion-thoughts. In an episode of anger we need to know all thoughts related to the event. These thoughts are not real; but they are connected with sensations, the bodily feelings of contraction. We need to observe where the muscles contract and where they don't. Some people get angry in their faces, some people get angry in their backs, some people get angry all over. The more we know — the stronger the observer is — the less mysterious these emotions are, and the less we tend to get caught by them.

There are several ways to practice. One is with sheer concentration (very common in Zen centers), in which we take a koan and push hard to break through. In this approach what we are really doing is pushing the false thought and emotion into hiding. Since they are not real, we suppose that it is OK to push them out of the way. And it's true that if we are very persistent and push on a koan long enough, we can sometimes break through temporarily to the wonder of a life that is free of ego. Another way, which is our practice here, is slowly to open ourselves to the wonder of what life is by meticulous attention to the anatomy of the present moment. Slowly, slowly we become more sophisticated and knowledgeable, so that (for example) we may know that when we dislike a person, the left corner of our mouth pulls down. In this approach everything in our life—the good and bad events, our excitement, our depression, our disappointment, our irritability—becomes grist for the mill. It's not that we seek out the struggles and problems; but a mature student almost welcomes them, because we gradually learn from experience that as this anatomy becomes clear, the freedom and compassion increase.

A third way of practice (which I view as poor) is to substitute a positive for a negative thought. For example: if we are angry we substitute a loving thought. Now this changed conditioning may make us feel better. But it doesn't stand up well to the pressures of life. And to substitute one conditioning for another is to miss the point of practice. The point is not that a positive emotion is better than a negative one, but that *all* thoughts and emotions are impermanent, changing, or (in Buddhist terms) empty. They have no reality whatsoever. Our only freedom is in knowing, from years of observation and experiencing, that all personally centered thoughts and emotions (and the actions born of them) are empty. They are empty; but if they are not seen as empty they can be harmful. When we realize this we can abandon them. When we do, very naturally we enter the space of wonder.

This space of wonder—entering into heaven—opens when we are no longer caught up in ourselves: when no longer "It is I," but "It is Thou." I am all things when there is no barrier. This is the life of compassion, and none of us lives such a life all the time. In the eye-gazing practice, in which we meditate while facing another person, when we can put aside our personal emotions and thoughts and truly look into another's eyes, we see the space of no-self. We see the wonder, and we see that this person *is* ourselves. This is marvelously healing, particularly for people in relationships who aren't getting along. We see for a second what another person is: they are no-self, as we are no-self, and we are both the wonder.

Some years ago in a workshop I did the eye-gazing exercise with a young woman who said her life had been shattered by the death of her father. She said that nothing she had done had given her any peace with this loss. For sixty minutes, we looked into each other's eyes. Because of zazen practice, I had enough power that it was easy for me to keep my gaze steady and unbroken. When she wavered, I could pull her back. At the end she started to cry. I wondered what was wrong, but then she said, "My father hasn't gone anywhere! I haven't lost him. It's fine, I'm at peace at last." She saw who she was and who her father was. Her father was not just a body that had disappeared. In the wonder, she was reconciled.

We can practice observing ourselves becoming angry: the arising thoughts, the bodily changes, the heat, the tension. Usually we don't see what is happening because when we are angry, we are identified with our desire to be "right." And to be honest, we aren't even interested in practice. It's very heady to be angry. When the anger is major we find it hard to practice with it. A useful practice is to work with all the smaller angers that occur everyday. When we can practice with those as they occur, we learn; then when the bigger uproars come that ordinarily would sweep us away, we don't get swept away so much. And over time we are caught in our anger less and less and less.

There is an old koan about a monk who went to his master and said, "I'm a very angry person, and I want you to help me." The master said, "Show me your anger." The monk said, "Well, right now I'm not angry. I can't show it to you." And the master said, "Then obviously it's not you, since sometimes it's not even there." Who we are has many faces, but these faces are not who we are.

I have been asked, "Isn't observing a dualistic practice? Because when we are observing, something is observing something else." But in fact it's not dualistic. The observer is empty. Instead of a separate observer, we should say there is just *observing*. There is no one that hears, there is just hearing. There is no one that sees, there is just seeing. But we don't quite grasp that. If we practice hard enough, however, we learn that not only is the observer empty, but that which is observed is also empty. At this point the observer (or witness) collapses. This is the final stage of practice; we don't need to worry about it. Why does the observer finally collapse? When nothing sees nothing, what do we have? Just the wonder of life. There is no one who is separated from anything. There is just life living itself: hearing, touching, seeing, smelling, thinking. That is the state of love or compassion: not "It is I," but "It is Thou."

So the way of practice that I've found to be the most effective is to increase the power of the observer. Whenever we get upset we have lost it. We can't get upset if we are observing, because the observer never gets upset. "Nothing" can't get upset. So if we can be the observer, we watch any drama with interest and affection, but without being upset. I've never met anyone who had completely become the observer. But there is a vast difference between someone who can be it most of the time and someone who can be it only rarely. The aim of practice is to increase that impersonal space. Although it sounds cold—and as a practice it *is* cold—it doesn't produce cold people. Quite the opposite. When we reach a stage where the witness is collapsing, we begin to know what life is. It's not some spooky thing, however; it just means that when I look at another person, I *look* at them; I don't add on ten thousand thoughts to what I am seeing. And that is the space of compassion. We don't have to try to find it. It's our natural state when ego is absent.

We have turned into very unnatural beings. But with all our difficulties, we have an opportunity open to us that no other animal has. A cat *is* the wonder; but the cat doesn't know that, it just lives it. But as human beings we have the capacity to realize it. As far as I know, we are the only creature on the face of the earth that has that capacity. Having been given this capacity—being made in the likeness of God—we should be endlessly grateful that we have the opportunity to realize what life is and who we are.

So we need to have patience—not just during sesshin, but every day of our lives—to face this challenging task: meticulously to observe all aspects of our life so that we can see their nature, until the observer sees nothing when it looks out except life as it is, in all its wonder. We all have such moments. After a sesshin, we may look at a flower and for a second there is no barrier. Our practice is to open our life like this more and more. That's what we are here on earth to do. All religious disciplines at bottom say the same thing: I and my Father are one. What is my Father? Not something other than myself, but just life itself: people, things, events, candles, grass, concrete, I and my Father are one. As we practice, we slowly expand this realization.

Sesshin is a training ground. I'm just as interested in what you will be doing two weeks from now when you find yourself in a crisis. Then will you understand how to practice? Observing your thoughts, experiencing your body instead of getting carried away by the fearful thoughts, feeling the contraction in your stomach as just tight muscles, grounding yourself in the midst of crisis. What makes life so frightening is that we let ourselves be carried away in the garbage of our whirling minds. We don't have to do that. Please sit well.

VI. IDEALS

Running in Place

I talk to many people, and I am saddened again and again that we don't see what our life is and what our practice is. We are confused about the basic core of practice, and we get sidetracked with all sorts of incorrect notions about it. The degree to which we're sidetracked or confused is the degree to which we suffer.

Practice can be stated very simply. It is moving from a life of hurting myself and others to a life of not hurting myself and others. That seems so simple—except when we substitute for real practice some idea that we should be different or better than we are, or that our lives should be different from the way they are. When we substitute our ideas about what *should* be (such notions as "I should not be angry, or confused, or unwilling") for our life as it truly is, then we're off base and our practice is barren.

Suppose we want to realize how a marathon runner feels: if we run two blocks, or two miles or five miles, we will know something about running those distances, but we won't yet know anything about running a marathon. We can recite theories about marathons; we can describe tables about the physiology of marathon runners; we can pile up endless information about marathon-running; but that doesn't mean we know what it is. We can only know when we are the one doing it. We only know our lives when we experience them directly, instead of dreaming about how they might be if we did this, or had that. This we can call running in place, being present as we are, right here and right now.

The first stage in practice is to recognize that we're not running in place, we're always thinking about how our lives might be (or how they once were). What is there in our life right now that we don't want to run in place with? Whatever is repetitive or dull or painful or miserable: we don't want to run in place with that. No indeed! The first stage in practice is to realize that we are rarely present: we're not experiencing life, we're thinking about it, conceptualizing it, having opinions about it. It is frightening to run in place. A major component of practice is to realize how this fear and unwillingness dominates us.

If we practice with patience and persistence, we enter the second stage. We slowly begin to be conscious of the ego barriers of our life: the thoughts, the emotions, the evasions, the manipulations, can now be observed and objectified more easily. This objectification is painful and revealing; but if we continue, the clouds obscuring the scenery become thinner.

And what is the crucial, healing third stage? It is the direct experiencing of whatever the scenery of our life is at any moment as we run in place. Is it simple? Yes. Is it easy? No.

I remember the Saturday morning when we delayed the morning sitting schedule for twenty minutes so that some of us could go a few blocks for the great opportunity of watching the San Diego marathon leaders race by. At 9:05, along they came. I was amazed by the flowing quality of the leader's movement: even though he was in the final five miles, he was simply gliding along. It was not hard to appreciate his running—but where is it that *we* have to run and practice? We must practice with ourselves as we are right now. To see a top-level runner is inspiring, but thinking that we should be like that now is not useful. We have to run where we are, we have to learn here and now, from how we are here and now.

We never grow by dreaming about a future wonderful state or by remembering past feats. We grow by being where we are and experiencing what our life is right now. We must experience our anger, our sorrow, our failure, our apprehension; they can all be our teachers, when we do not separate ourselves from them. When we escape from what is given, we cannot learn, we cannot grow. That's not hard to understand, just hard to do. Those who persist, however, will be those who will grow in understanding and compassion. How long is such practice required? Forever.

Aspiration and Expectation

Aspiration is a basic element of our practice. We could say that the whole practice of Zen comes out of our aspiration; without it nothing can happen. At the same time we hear that we should practice without any expectation. It sounds contradictory because we quite often confuse aspiration and expectation.

Aspiration, in the context of practice, is nothing but our own true nature seeking to realize and express itself. Intrinsically we are all Buddhas, but our Buddha nature is covered up. Aspiration is the key to practice because, without it, our Buddha nature is like a beautiful car: until someone gets in the driver's seat and turns on the ignition it is useless. When we begin to practice our aspiration may be very small; but as we continue in the practice aspiration grows. After six months of practice one's aspiration is different from what it was when just beginning practice, and after ten years it will be different than after six months. It is always changing in its outward form, yet essentially it is always the same. As long as we live it will continue to grow.

One sure clue as to whether we're being motivated by aspiration or expectation is that aspiration is always satisfying; it may not be pleasant, but it is always satisfying. Expectation, on the other hand, is always unsatisfying, because it comes from our little minds, our egos. Starting way back in childhood, we live our lives looking for satisfaction outside ourselves. We look for some way to conceal the basic fear that something is missing from our lives. We go from one thing to another trying to fill up the hole we think is there.

There are many ways by which we try to hide from our dissatisfaction. One way, for instance, is by struggling to achieve something. Now achievement in itself is a natural thing, and it's important that we learn to run our lives well. But as long as we look outside ourselves for some reward in the future, we are bound to be disappointed in our expectations. Life takes care of that very well; it has a way of disappointing us, efficiently and regularly.

Generally we look at life in terms of two questions: "Will I get something out of this?" or "Will this hurt me?" We may seem serene, but under that surface of serenity these two questions bubble and boil. We come to a spiritual practice like Zen trying to find the peace and satisfaction that has so far eluded us, and what do we do? We take the same habits that we lived with all our lives and put our practice into that same framework. We set up one goal after another, continuing this lifelong habit of running after something: "I wonder how many koans I can pass this sesshin"; or "I've been sitting longer than she, but she seems to be progressing more quickly"; or "My zazen was so wonderful yesterday—if only I could get it back again." In one way or another, our approach to practice is based on the same struggle to achieve something: to be recognized by our peers, to be important in the Zen world, to find a safe hole to hide in. We're doing the same thing we've always done; we're expecting something (in this case Zen practice) to give us satisfaction and safety.

Dōgen Zenji said, "To look for the Buddha dharma outside of yourself is like putting a devil on top of yourself." Master Rinzai said, "Place no head above your own." That is, to look outside of ourselves for true peace and satisfaction is hopeless.

It's important that we continually examine ourselves and see where it is that we're looking and what it is that we're looking for. What are you looking for outside of yourself? What is it that you think is going to do it? Position? Relationships? Passing koans? Over and over again the Zen masters say to place no head above your own, and add nothing extra to your life.

Each moment, as it is, is complete and full in itself. Seeing this, no matter what arises in each moment, we can let it be. Right now, what is your moment? Happiness? Anxiety? Pleasure? Discouragement? Up and down we go, but each moment is exactly what each moment is. Our practice, our aspiration, is to be that moment and let it be what it is. If you are afraid, just be fear, and right there you are fearless.

There's a story of three people who are watching a monk standing on top of a hill. After they watch him for a while, one of the three says, "He must be a shepherd looking for a sheep he's lost." The second person says, "No, he's not looking around. I think he must be waiting for a friend." And the third person says, "He's probably a monk. I'll bet he's meditating." They begin arguing over what this monk is doing, and eventually, to settle the squabble, they climb up the hill and approach him. "Are you looking for a sheep?" "No, I don't have any sheep to look for." "Oh, then you must be waiting for a friend." "No, I'm not waiting for anyone." "Well, then you must be meditating." "Well, no. I'm just standing here. I'm not doing anything at all."

It's very difficult for us to conceive of someone just standing and doing nothing because we are always frantically trying to get somewhere to do *something*. It's impossible to move outside of this moment; nevertheless, we habitually try to. We bring this same attitude to Zen practice: "I know Buddha-nature must be out there somewhere. If I look hard enough and sit hard enough, I'll find it eventually!" But seeing Buddha-nature requires that we drop all that and completely be each moment, so that whatever activity we are engaged in—whether we're looking for a lost sheep, or waiting for a friend, or meditating—we are standing right here, right now, doing nothing at all.

If we try to make ourselves calm and wise and wonderfully enlightened through Zen practice, we're not going to understand. Each moment, just as it is, *is* the sudden manifestation of absolute truth. And if we practice with the aspiration just to be the present moment, our lives will gradually transform and grow wonderfully. At various times we'll have sudden insights; but what's most important is just to practice moment by moment by moment with deep aspiration.

When we are willing just to be here, exactly as we are, life is always OK: feeling good is OK, feeling bad is OK; if things go well it's OK, if things go badly it's OK. The emotional upsets we experience are problems because we don't want things to be the way they are. We all have expectations, but as practice develops those expectations gradually shrivel up and, like a withered leaf, just blow away. More and more we are left with what is right here, right now. This may seem frightening, because our expecting mind wants life to turn out a certain way: we want to feel good, we don't want to be confused, we don't want to get upset—each of us has our own list.

But when you're tired after work, that's the tired Buddha; when your legs hurt during zazen, that's the hurting Buddha; when you're disappointed with some aspect of yourself, that's the disappointed Buddha. That's it!

When we have aspiration we look at things in a completely different way than we do when we have expectation. We have the courage to stay in this moment since, in fact, this moment is all we ever have. If the mind wanders off into expectations, having aspiration means gently returning it to the present moment. The mind will wander off all the time, and when it does, simply return to the moment without worrying or getting excited. Samadhi, centeredness, and wholeness will develop naturally and inevitably from this kind of practice, and aspiration itself will grow deeper and clearer.

Seeing Through the Superstructure

Suppose that we talk about our life as though it were a house, and we live in this house, and life goes along as it goes along. We have our stormy days, our nice days; sometimes the house needs a little paint. And all the drama that goes on within the house between those who live there just goes on. We may be sick or well. We may be happy or unhappy. It's like this for most of us. We just live our life, we live in a house or an apartment and things take place as they take place. But—and this is when practice becomes important—we have this house but it's as though it were encased in another house. It's as though we took a strawberry and we dipped it in chocolate; so we have our strawberry and it's covered with coating. We have our perfectly nice house, and on top of that house we have another house, encasing this basic house in which we live.

Yet our life (the house) as we live it is perfectly fine. We don't usually think so, but there's nothing wrong with our life. Just as it is. But we erect, right on top of the house we have, an extra one. And if we haven't looked closely at what we've added on, the extra layer can be very thick and dark. The house we live in will then seem dim and confined because we've covered it over with something heavy. That covering can seem impenetrable, frightening, depressing. The biggest error we make in our life and our practice is to think that this house we're living in—which is our life just as it is, with all of its problems, all of its ups and downs—has something wrong with it. And because we think that, we get busy. We've been busy most the years of our life adding that extra structure.

Zen practice is first of all to see that we have done that; and then it is about seeing what the superstructure is—how it operates, what it's made of, what we have to do with it or what we don't have to do with it. Usually we think, "It's unpleasant; I must get rid of it." No, I don't think that's the way. Essentially, this extra structure covering our life has no reality. It has come to be there because of the misuse of our minds. It's not a question of getting rid of it, since it has no reality; but it *is* a question of seeing its nature. And as we see its nature, instead of it being so thick and dark, the covering becomes more transparent: we see through it. Enlightenment (bringing in more light) is what happens in practice. Actually we're not getting rid of a structure, we're seeing through it as the dream it is, and as we realize its true nature its whole function in our life weakens; and at the same time we can see more accurately what is going on in our daily life. It's as if we have to go full circle. Our life is always all right. There's nothing wrong with it. Even if we have horrendous problems, it's just our life. But since we refuse to accept life as it is, because of our preference for things that are pleasurable, we pick and choose from life. Said another way, we have no intention of settling for life as it is when it does not suit us.

Each one of you sitting here has some set of events that you're simply not willing to have be your life. "It's not that! It couldn't be that!" For example: when I was a teenager, if I didn't have a date for Saturday night, *no way* was it OK for me not to have a date. I'd put all sorts of extras on top of the dismal fact of no date: "Something's wrong; I should change my hair, I need a different color of nail polish. I need . . . I need . . ." That's a silly example, of course. But in the major traumas in our life we do the same thing. Because of our unwillingness for life to be the way it is, we always add something extra. There is no one here who doesn't do that. Nobody. As long as we live, we will probably always have at least a thin overcoating on the essential structure of our life. But how much is the question.

Zen practice isn't about a special place or a special peace, or something other than being with our life just as it is. It's one of the hardest things for people to get: that my very difficulties in this very moment *are* the perfection. "What do you mean, they're the perfection? I'm gonna practice and get rid of them!" No, we don't have to rid of them, but we must see their nature. The structure becomes thinner (or seems thinner); it gets lighter and occasionally we may crack a hole right through it. Occasionally. So one thing I want you to do is to identify for yourself what it is in your life right now that you're not willing to have be as it is. It could be troubles with your partner, it could be unemployment, it could be disappointment with some goal that has not been reached. Even if what is happening is fearful and distressing, it's fine. It's very difficult to get that. Strong practice is needed to make even a dent in our habitual way of viewing life. It's hard to get that we don't have to get rid of the calamity. The calamity is fine. You don't have to like it, but it's fine.

The first step in practice is to realize that we have erected this superstructure. And as we do zazen (particularly as we label our thoughts) we begin to recognize that we're almost never just living our life as it is. Our lives are lost in our self-centered thoughts, the superstructure. (We are presuming that we *want* to see through this superstructure. Some people simply don't. And that's OK too. Not everyone should be doing a practice like Zen. It's demanding, it's disillusioning. It can seem forbidding when we are new to it. That's just one side of it. The other side is that life becomes vastly more fulfilling as we practice. The two sides go together.) So the process of practice is, first of all, to have an awareness, maybe dim at first, of what we have erected; and the second step is to practice. Liberation is to see through this unreal superstructure that we've built. Without it, life just goes along as it goes along with no obstacle. Does that make sense? Sounds crazy, doesn't it?

Let's realize that our ideals *are* the superstructure. When we are attached to the way we think we should be or the way we think anyone else should be, we can have very little appreciation of life as it is. Practice must shatter our false ideals. So we're stating a fact that for most people is unacceptable. Right now, look at your practice and see if you want to do it. After we've been sitting a while, what comes up is, "I don't want to do this! I don't want to do it at all!" But that's part of practice too!

To look at this structure we have built is a subtle, demanding process. The secret is, we like that unreal structure a lot better than we like our real life. People have been known to kill themselves rather than demolish their structure. They will actually give up their physical life before they will give up their attachment to their dream. Not uncommon at all. But whether or not we commit physical suicide, if our attachment to our dream remains unquestioned and untouched, we are killing ourselves, because our true life goes by almost unnoticed. We're deadened by the ideals of how we think we should be and the way we think everybody else should be. It's a disaster. And the reason we don't understand that it's a disaster is because the dream can be very comfortable, very seductive. Ordinarily we think a disaster is an event like the sinking of the *Titanic*. But when we are lost in our ideals and our fantasies, pleasurable as they may be, this *is* a disaster. We die.

Another point. My daughter and I were talking about a man who was doing vicious things, and I muttered, "He should be more conscious of what he is doing." She laughed and said, "Mom, if you're unconscious, the nature of being unconscious is what? Just to be unconscious." And she was right: to be unconscious means you *don't* see what you are doing. So one of the problems in practice is that we are all to some degree unconscious, and not very willing to be conscious. How to cut through that? Part of that is my job. Most of it is yours. I remember a senior student years ago who had just given a lovely talk on giving and compassion. A day later I watched him when the call was made to fill the line to see the teacher. This man practically clawed his way to the front of the line—quite unconscious of his selfishness. Until we see what we're doing, we will do it. So in practice one of our tasks is to keep upping this ability to see. Very tricky, since we don't have a great deal of interest in seeing anyway!

Discipline has a connotation for some of us of forcing ourselves to do something. But discipline is simply bringing all the light we can summon to bear on our practice, so that we can see a little bit more. Discipline can be formal, as in the zendo, or informal, as in our daily life. Disciplined students are those who in their everyday activities constantly try to find means of waking themselves up.

The question is always the same: in this moment, what do we see and what *don't* we see? If we're practicing well, one day we will see something we have never seen before. Then we can work with it. Practice is to keep that subtle pressure operating from morning till night. As we do the superstructure will begin to lighten and we can see more clearly our life as it is.

I'm talking here about the general course of practice, and such talks may emphasize something too much and something else too little. It's inevitable. Questions can help to clarify the talk.

STUDENT: Yes, there are two of me over here and we get confused when you give talks like this. This first one of me has a lot of ideals . . .

JOKO: Well, that's what we want to demolish.

STUDENT: Are you saying that I shouldn't be involved in service organizations?

JOKO: Oh, of course not!

STUDENT: But that's an ideal!

JOKO: No, no, no . . . it's not an ideal, you just do it. But recognize any idealistic thoughts you add on to what you do. If someone's dying of hunger in the front yard, we certainly don't question what to do. We go and get some food. But then we may notice how nice we are to do this. That's what we add; that's the superstructure. There's the action itself, and then there's the superstructure. By all means, *do*. The most efficient way to wear out the superstructure is to keep doing all the nonsense that we're always doing, but to do it with as much awareness as we can possibly muster. Then we see more.

STUDENT: Well, that's one-half of me. The other half is unemployed and depressed and kind of hungry, and there are people depending on me. And what I hear you saying is, I should just appreciate my hunger and my unemployment and maybe I shouldn't look for a job?

JOKO: No, no. By no means! If you're out of a job exert yourself to get a job. Or if you're sick do whatever you can to get better. But it's what you add on to these basic actions which is the superstructure. It could be, "I'm such a hopeless person, nobody would ever want to employ me anyway!" That's superstructure. Being unemployed means to look at your job possibilities in the present job market and if necessary to get some training to increase your skills. But what do we always add on to the basic facts of a situation?

STUDENT: I've been looking at my parents' lives and my interaction with them. In certain areas they seem to be weak, and I seem to have some issues with that. Psychologists say that the first five years of one's life are so impressionable that they form the basis for one's life. Would you comment on that?

JOKO: Well, there is the absolute point of view and the relative point of view. From the relative point of view we have a history. Much has happened to each one of us, and we are as we are at least partly because of our history. But in another sense we have *no* history. Zen practice is to see through our desire to cling to our history and to reasons (thoughts) for *why* we are as we are, instead of working with the reality of *what* we are. There are many kinds of therapy. But any therapy that leads you to feel that your life is terrible because of what someone did to you is at least incomplete, because each of us has had a lot done to us, right? But our responsibility is always right here, right now, to experience the reality of our life as it is. And eventually to blame no one. If we blame anyone we know we're caught; we can be sure of that.

STUDENT: How do you know?

JOKO: How do I know what?

STUDENT: How do you know all this?

JOKO: I wouldn't say I know it . . . I think it's just obvious from years of sitting. And I'm not telling you to believe it. I don't want anyone here to believe what I'm saying—I want you to work with your own experience. And then see for yourself what's true for you. But what in particular do you question about what I was saying?

STUDENT: Maybe I'm questioning my willingness to believe you.

JOKO: But I don't want you to believe me! I want you to practice! We're almost like scientists working with our own life. If we are observant then we see for ourselves whether or not the experiment works. If we practice with our life and the superstructure lightens, then we'll know for ourself. Certain religions just say, "Believe." Belief has no part in what we are doing here. I don't want anyone to believe me. But it's not going to hurt you to practice. There's nothing I'm telling you to do that would be harmful to you.

STUDENT: My question is mixed with his. It seems that to do this practice we have to have a lot of faith in ourselves. This is the way it feels to me.

JOKO: Well, call it faith if you want. I don't think you would be here if you didn't feel that practice would be useful to you. And that's faith in a way.

STUDENT: In my judgment I feel it's important to know what happened to me in childhood . . .

JOKO: I didn't say it doesn't have some usefulness. But your experience in this moment embraces your whole life, including the past, and it depends on whether or not you know how to experience that — *really* experience it. See, we talk a lot about being our experience. But to do that is not easy and we do it very infrequently. It's one thing to give talks about experiencing what is, but it's difficult to do it so we avoid it. But when we practice well, our life — past and present — will resolve itself. Little by little.

STUDENT: What place do prayer and affirmation have in Zen practice?

JOKO: Prayer and zazen are the same thing; there's no difference. Affirmations I would avoid, because an affirmation (like "I'm really a healthy person") may produce temporary feelings of well-being, but it doesn't acknowledge the present reality — which might be that I'm sick.

STUDENT: How about all the evil forces around us that seem to be getting stronger?

JOKO: I don't think that there are evil forces around us. I think that there are evil acts being done, but that's quite different. If someone is hurting a child you certainly want to stop the action; but to condemn the person doing it is as evil is unsound practice. We should oppose evil actions, but people — no. Otherwise we'll go around judging and condemning everyone, including ourselves.

STUDENT: By the same token, then, you can't call someone good either.

JOKO: Right. Essentially, in Zen terms we are "no-thing" . . . we just are doing what we're doing. But if we see the unreality of this superstructure, we tend to do good. When there is no separation between ourselves and others, naturally we do good. Our basic nature is to do good.

STUDENT: That's our action.

JOKO: Yes, we just do it naturally. If we're not separated from others by our self-centered thoughts of greed, anger, and ignorance, we will do good. We don't have to force ourselves to do that. It's our natural state.

Prisoners of Fear

We all know the picture of the important executive working until ten o'clock at night, answering the phone, grabbing a sandwich on the run. His poor body is being short-changed. He thinks his frantic efforts are essential for "the good life"; he fails to see that desire is running his life—and it runs all of ours, too. Because we are controlled by our desires, we have only a dim awareness of the basic truth of our existence.

Most people who don't do any kind of practice are pretty selfish; they are caught up in desires: to be important, to possess this or that, to be rich, famous. Of course that's true of all of us, to a degree. But as we practice we begin to suspect that our life is not working quite the way the TV commercials say it will. Television advertisements suggest that if you have the newest hair spray and makeup and garage door opener, you life is going to be great. Right? Well, most of us find that isn't true. And as we see that, we begin to see that the way we live isn't working. The selfish greed which runs our lives is not working.

Then we begin a second stage: "Well, if it doesn't work to be selfish, then I'm going to be unselfish." Most religious practices (and some Zen practices, I'm sorry to say) are about unselfish-ness. Seeing our meanness, our unkindness, we decide to pursue a new desire: to be kind, to be good, to be patient. Guilt goes with this desire, like a sort of baby brother: when we don't fulfill our picture of how we should be, we feel guilty. We're still trying to be something we're not. We're trying to figure out how to be different than we are. When we can't fulfill our ideals, we build guilt and depression. In our practice we swing through both of these stages. We see that we are mean, greedy, violent, selfish, ambitious. And so we form a new ambition, to be unselfish. "I shouldn't have those thoughts. I've been sitting for such a long time; why am I still greedy and mean? I should be better than that by now." We're all doing that. A lot of religious practice mistakenly aims at trying to produce a good person who doesn't do or think bad things. Some Zen centers are caught up in this approach, too. It leads to a kind of arrogance and self-righteousness because if you are doing it *right*, what about all the others who don't know the truth and are *not* doing it right? I've had people say to me, "Our ses-shins begin at 3 A.M. When do yours begin? At 4:15? Oh . . ." So this second stage has a lot of arrogance in it. Guilt has a lot of arrogance in it. I'm not saying it's bad to be arrogant; that's just what we are if we don't see.

Yet we make a tremendous effort to be good. I've heard people say, "Well, I was just out of sesshin and someone cut me off on the freeway—and what do you know, I was angry. What a poor student I am . . ." We all do that. See, all wanting—especially wanting to be a certain way—is centered on ego and fear. "If I can be perfect, if I can be realized or enlightened, I will take care of the fear." Do you see the desire there? There's a tremendous desire to move away from what I am, into an ideal. Some people don't care about enlight-enment; but they may feel, "I shouldn't yell at my spouse." Of course you shouldn't yell at your spouse; but the effort to be a person who doesn't yell at your spouse just increases the tension.

To move from being selfish and greedy to trying not to be that way is like taking down all the drab and ugly pictures in your room and putting up pretty pictures. But if that room is a prison cell, you've changed the decorations and they look a little better; but still the freedom you want isn't there; you're still imprisoned in the same room. Changing the pictures on the wall from greed, anger, and ignorance into ideals (that we should not be greedy, angry, or ignorant) improves the decoration, perhaps—but leaves us without freedom.

I'm reminded of an old story about a king who wanted the wisest man among his subjects to be his prime minister. When the search finally was narrowed to three men the king put them to a supreme test: he placed them in a room in his palace and installed an ingenious lock in the door. The candidates were told that the first person to open the door would be appointed prime minister. So two of them started to work out complicated mathematical formulas to discover the proper lock combination. But the third man just sat in his chair for a time—and then, without bothering to put pen to paper, he got up, walked to the door, and turned the knob—and the door opened. It had been unlocked the whole time. What is the point of that story? The prison cell we live in, whose walls we are frantically redecorating, is not a prison cell. In fact the door has never been locked. There is no lock. We don't need to sit in our cells and struggle for freedom by frantically trying to change ourselves—because we are already free.

Merely to say this doesn't solve the problem for us, of course. How can we realize this fact of freedom? We've said that being selfish and having a desire to be unselfish are both based on fear. Even the desire to be wise, to be perfect, is based on fear. We wouldn't chase the desire if we saw that we were already free. So our practice always comes back to the same thing: how to see more clearly, how not to go down blind alleys, such as that of trying to be unselfish. Instead of going from unconscious selfishness to conscious unselfishness, what we need to do is to see the foolishness of the second stage—or, if we play around in it, at least to be aware that we are doing so. What we need is to go to the *third* stage, which is . . . what?

Initially we must tear the first two stages apart. We do this by becoming the witness. Instead of saying, "I should not be impatient," we observe ourselves being impatient. We stand back and watch. We see the truth of our impatience. The truth is certainly not some mental picture of being nice and patient; in creating that picture we just bury the irritation and anger, which will pop out later. What *is* the truth of any moment of upset, when we are impatient, jealous, or depressed? When we start working like this—which means to really observe our minds—we see that they are constantly spinning dreams of how we should or shouldn't be or how someone else should or shouldn't be; of how we've been in the past, and how we're going to be in the future; of how we can arrange matters to get what we desire.

When we step back and become a patient and persistent witness, we begin to understand that neither of the two stages does ourselves or anyone else any good. Only then can we—without even trying—slip into stage three, which means simply to experience the truth of that moment of impatience, the very fact of just feeling impatient. When we can do that we have slipped out of the duality that says there is me and there is a way I should be—and we return to ourselves as we are. And when we experience ourselves as we are—since the only thing that is holding impatience in place is our thoughts—the impatience begins to resolve itself.

So our practice is about making fear conscious, instead of running around inside our cell of fear, trying to make it look better and feel better. All of our efforts in life are these escaping endeavors: we try to escape the suffering, escape the pain of what we are. Even feeling guilty is an escape. The truth of any moment is always being just as we are. And that means to experience our unkindness when we are unkind. We don't like to do that. We like to think of ourselves as kind people. But often we're not.

When we experience ourselves as we are, then out of that death of the ego, out of that withering, the flower blooms. On a withered tree, the flower blooms—a wonderful line from *Shōyō Rōku*. A flower blooms, not on a decorated tree, but on a withered tree. When we back away from our ideals and investigate them by being the witness, then we back into what we are, which is the intelligence of life itself.

How does the process we're talking about relate to enlightenment? When we back out of unreality by witnessing it, we see it for what it is; we fall into reality. Maybe at first we see it only a second at a time, but over time the percentage goes up. And when we can spend over ninety percent of our time being with life as it is, we're going to see *what* life is. We *are* life, then. When we are anything, we know what it is. We're like the earnest fish that spent its lifetime swimming from teacher to teacher. The fish wanted to know what the ocean was. And some teachers told him, "Well, you have to try very hard to be a good fish. This is a tremendous area that you're investigating. And you have to meditate for long hours, and you have to punish yourself and you have to really really try to be a good fish." But the fish at last came to one teacher and asked, "What's the great ocean? What's the great ocean?" And the teacher simply laughed.

Great Expectations

Two book titles recently jumped into my mind. The first was *Great Expectations* by Charles Dickens and the second was *Paradise Lost* by John Milton. There is an intimate connection between them. What is it?

We are all seeking paradise, enlightenment, or whatever name you choose to use. It seems to us that paradise is lost: "There's not much of it in *my* life," most people would say. We want this "paradise," this "enlightenment"; we are desperate for it. We are here to seek for it. But where is it? What is it?

We come to sesshin with great expectations. And we struggle, we search, we hope. And some of us even expect. The human game continues. We have, if not great expectations, some hope that sometime paradise is going to appear to us.

Now if we don't know what paradise is, we know for sure what it isn't. We are certain that paradise is not about being miserable. Paradise is not failing at something. Paradise is not being criticized or humiliated or punished in any way. It's not physical pain. It's not making mistakes. It's not losing my partner or my friend or my child. Paradise just couldn't be confusion or depression. And it's not about being lonely, or working when worn out or sick. We have definite lists of what paradise is *not*. But if it's not these states, then what is it?

Is it having more money, more security? Is it having dominance or power or fame or recognition from others? Is paradise being surrounded by people, being supported and loved? Is it having more peace and quiet, more time to think about the meaning of life? Is it any of these? Or is it not?

Some people here have "made it" into the second list. They've got some of these things, a little of "the good life." And yet no matter what we have, once we have it . . . "Oh, is *this* it? But no, this isn't it either." Where is it? We never seem to quite catch it. It's like chasing a mirage: when we arrive, it disappears.

It's interesting that some people, as they near their death, finally see or realize what they had never seen or realized up to that point. And having that realization they die peacefully, even with joy, in paradise at last. What have they seen? What have they found?

Remember the tale of the man being chased by a tiger? Facing death before him and behind him, he eats a strawberry, exclaiming, "It's so delicious!"—because he knows that for him, it is his last act.

Now let's return to our first list—what paradise isn't—and give it a fresh slant. "I'm so miserable. And it's so delicious!" "I really failed. And it's so delicious!" "I've never been so humiliated in my life. And it's so delicious!" "I'm so lonely. And it's so delicious!" When we thoroughly understand this, any circumstance of life is paradise itself.

Now let's turn to some words of Dōgen Zenji. He once said, "Let go of and forget your body and mind. Throw your life into the abode of the Buddha, living by being moved and led by the Buddha.

When you do this without relying on your own physical and mental power, you become released from life and death and become a Buddha. This is the Truth. Do not search for the Truth anywhere else."

"Let go of and forget your body and mind." What is meant? "Throw your life into the abode of the Buddha." What is the abode of the Buddha? He refers to human error in his first words: "Let go of and forget your body and mind." Instead of referring everything to the comfort, protection, and pleasure of body and mind – which we do – he asks us to "throw your life into the abode of the Buddha." But where is the abode of the Buddha? Where are we to throw our life?

Since Buddha is none other than this absolute moment of life (which is not the past or the present or the future), he is saying that this very moment *is* the abode of the Buddha, enlightenment, paradise. It is nothing but the life of this very moment. Whether we are miserable or happy, a failure or a success, there is *nothing* we experience which is not the abode of the Buddha. "Throw your life into the abode of the Buddha, living by being moved and led by it." What is meant?

We cannot live without being this moment, because that's what our life is. Being led by it is to see it, feel it, taste it, touch it, experience it, and then let it dictate what is to be done. He says that when you do this without relying on your own physical and mental power – that is, without your personal opinions as to how things should be – you become released from both life and death and become a Buddha. Why? Why do you become a Buddha? Because you *are* a Buddha. You *are* this moment of life. You can't be anything else.

When we sit, when we live our daily schedules, we are in the abode of the Buddha. Where else could we be? Every moment of zazen, painful, peaceful, boring, is what? Paradise, nirvana, the abode of the Buddha. Yet we come with great expectations to sesshin to attempt to find it! Where is it? When you leave here, where is it? The abode of the Buddha is your body and mind's direct experience. Not something else, somewhere else. Dōgen Zenji said, "This is the Truth. Do not search for the Truth anywhere else." Where can you search?

There is no paradise lost, none to be regained. Why? Because you cannot avoid this moment. You may not be awake to it, but it is always here. You cannot avoid paradise. You can only avoid seeing it.

When people know their death is very close, what is the element that often disappears? What disappears is the hope that life will turn out the way they want it to. Then they can see that the strawberry is "so delicious"—because that's all there is, this very moment.

Wisdom is to see that there is nothing to search for. If you live with a difficult person, that's nirvana. Perfect. If you're miserable, that's it. And I'm not saying to be passive, not to take action; then you would be trying to hold nirvana as a fixed state. It's never fixed, but always changing. There is no implication of "doing nothing." But deeds done that are born of this understanding are free of anger and judgment. No expectation, just pure and compassionate action.

Sesshin is often a battle with the fact that we absolutely don't want our experience to be the way it is. We definitely don't feel it is the enlightened state. But patiently sitting through it and turning away from all such concepts—"It's hard, it's wonderful, it's boring, this shouldn't be happening to me"—enable us in time to realize the Truth of our lives. The first day of sesshin is all about the first list; the mind racing with all of the problems current in our lives, our desires, our frustrations; all compounded with the fatigue of the first day and often with some physical discomfort. All of our pet ideas are assaulted in sesshin.

And always we are seeking a way around such troubles into the elusive paradise. But again Dōgen Zenji's words, "Let go of your body and mind," remind us just to maintain clear awareness of all conditions of body and mind, noting our desire to seek pleasure and avoid pain. But both are this very moment. So he says, "Throw your life into the abode of the Buddha." Throw your life; be this very moment; cease to judge it, escape it, analyze it, just be it. He says, "This is the Truth. Do not search for the Truth any-where else." Why? Why can't we search for it somewhere else? There's no place else to search because there is nothing that ever happens except when? *Right here. Right now.* And it is our very nature, enlightenment itself. Can we wake up and look?

VII. BOUNDARIES

The Razor's Edge

We human beings all think there is something to accomplish, something to realize, some place we have to get to. And this very illusion, which is born out of having a human mind, is the problem. Life is actually a very simple matter. At any given moment in time we hear, we see, we smell, we touch, we think. In other words there is sensory input; we interpret that input, and everything appears.

When we are embedded in life there is simply seeing, hearing, smelling, touching, thinking (and I don't mean self-centered thinking). When we live this way there is no problem; there couldn't be. We are just *that*. There is life and we are embedded in it; we are not separate from life. We just are what life is because we are being what life is; we hear, we think, we see, we smell, and so on. We are embedded in life and there is no problem; life flows along. There is nothing to realize because when we are life itself, we have no questions about life. But that isn't the way our lives are—and so we have plenty of questions.

When we aren't into our personal mischief, life is a seamless whole in which we are so embedded that there is no problem. But we don't always feel embedded because—while life is *just* life—when it seems to threaten our personal viewpoint we become upset, and withdraw from it. For instance, something happens that we don't like, or somebody does something to us we don't like, or our partner isn't the way we like: there are a million things that can upset human beings. They are based on the fact that suddenly life isn't just life (seeing, hearing, touching, smelling, thinking) anymore; we have separated ourselves and broken the seamless whole because we feel threatened. Now life is *over there*, and I am *over here* thinking about it. I'm not embedded in it anymore; the painful event has happened *over there* and I want to think about it *over here* so I can figure a way out of my suffering.

So now we have split life into two divisions—over here and over there. In the Bible this is called "being banished from the Garden of Eden." The Garden of Eden is a life of unbroken simplicity. We all chance upon it now and then. Sometimes after sesshin this simplicity is very obvious, and for a while we know that life is not a problem.

But most of the time we have an illusion that life over there is presenting us with a problem over here. The seamless unity is split (or seems to be). And so we have a life harried by questions: "Who am I? What is life? How can I fix it so I can feel better?" We seem surrounded by people and events that we must control and fix because we feel separate. When we begin to analyze life, think about it, fuss and worry about it, try to be one with it, we get into all sorts of artificial solutions—when the fact of the matter is that from the very beginning, there is nothing that needs to be solved. But we can't see this perfect unity because our separateness veils it from us. Our life is perfect? No one believes *that*!

So there is life in which we truly are embedded (since all that we are *is* thinking, seeing, hearing, smelling, touching and we add on our self-centered thoughts about "how it doesn't suit me." Then we no longer can be aware of our unity with life. We've added something (our personal reaction) and when we do that, anxiety and tension begin. And we do this addition about every five minutes. Not a pretty picture . . .

Now what do I mean by the razor's edge? What we have to do to join together these seemingly separate divisions of life is to walk the razor's edge; then they come together. But what *is* the razor's edge?

Practice is about understanding the razor's edge and how to work with it. Always we have an illusion of being separate, which we have created. When we're threatened or when life doesn't please us, we start worrying, we start thinking about a possible solution. And without exception there is no person who doesn't do this. We dislike being with life as it is because that can include suffering, and that is not acceptable to us. Whether it's a serious illness or a minor criticism or being lonely or disappointed—that is not acceptable to us. We have no intention of putting up with that or just being that if we can possibly avoid it. We want to fix the problem, solve it, get rid of it. That is when we need to understand the practice of walking the razor's edge. The point at which we need to understand it is whenever we begin to be upset (angry, irritated, resentful, jealous).

First, we need to *know* we're upset. Many people don't even know this when it happens. So step number one is, be aware that upset is taking place. When we do zazen and begin to know our minds and our reactions, we begin to be aware that yes, we are upset.

That's the first step, but it's not the razor's edge. We're still separate, but now we know it. How do we bring our separated life together? To walk the razor's edge is to do that; we have once again to be what we basically are, which is seeing, touching, hearing, smelling; we have to experience whatever our life is, right this second. If we're upset we have to experience being upset. If we're frightened, we have to experience being frightened. If we're jealous we have to experience being jealous. And such experiencing is physical; it has nothing to do with the thoughts going on about the upset.

When we are experiencing nonverbally we are walking the razor's edge—we are the present moment. When we walk the edge the agonizing states of separateness are pulled together, and we experience perhaps not happiness but joy. Understanding the razor's edge (and not just understanding it, but doing it) is what Zen practice is. The reason it's difficult is that we don't want to do it. We know we don't want to do it. We want to escape from it.

If I feel that I've been hurt by you, I want to stay with my thoughts about the hurt. I want to increase my separation; it feels good to be consumed by those fiery, self-righteous thoughts. By thinking, I try to avoid feeling the pain. The more sophisticated my practice becomes, the more quickly I see this trap and return to experiencing the pain, the razor's edge. And where I might once have stayed upset for two years, the upset shrinks to two months, two weeks, two minutes. Eventually I can experience an upset as it happens and stay right on the razor's edge.

In fact the enlightened life is simply being able to walk that edge all the time. And while I don't know of anyone who can always do this, certainly after years of practice we can do it much of the time. It is joy to walk that edge.

Still I want to repeat: it is necessary to acknowledge that most of the time we want nothing to do with that edge; we want to stay separate. We want the sterile satisfaction of wallowing in "I am right." That's a poor satisfaction, of course, but still we will usually settle for a diminished life rather than experience life as it is when that seems painful and distasteful.

All troublesome relationships at home and work are born of the desire to stay separate. By this strategy we hope to be a separate person who really exists, who is important. When we walk the razor's edge we're not important; we're no-self, embedded in life. This we fear—even though life as no-self is pure joy. Our fear drives us to stay over here in our lonely self-righteousness. The paradox: only in walking the razor's edge, in experiencing the fear directly, can we know what it is to have no fear.

Now I realize we can't see this all at once or do it all at once. Sometimes we jump onto the razor's edge and then hop off, like water dropped on a sizzling frying pan. That may be all we can do at first, and that's fine. But the more we practice, the more comfortable we become there. We find it's the only place where we are at peace. So many people come to the Center and say, "I want to be at peace." Yet there may be little understanding of how peace is to be found. Walking the razor's edge is it. No one wants to hear that. We want somebody who will take our fear away or promise us happiness. No one wants to hear the truth, and we won't hear it until we are ready to hear it.

On the razor's edge, embedded in life, there is no "me" and no "you." This kind of practice benefits all sentient beings and that, of course, is what Zen practice is about . . . my life and your life growing in wisdom and compassion.

So I want to encourage you to understand, difficult though it may be. First we have to understand with the intellect: we must know intellectually what practice is. Then we need to develop through practice an acute awareness of when we are separating ourselves from our life. The knowing develops from the base of daily zazen, from many sesshins, and with the effort to remain aware in all encounters from morning till night. Since we are most unwilling to know about the razor's edge, this wisdom is not going to be presented on a platter to us; we have to earn it. But if we are patient our vision will become clearer and then we will see the jewel of that life, beginning to shine. Of course the jewel is always shining, but it is invisible to those who do not know how to see. To see, we must walk the razor's edge. We protest, "No! No way! Forget it! It's a nice title for a book, but I don't want it in my life." Is that true? I think not. Basically we *do* want peace and joy.

STUDENT: Please talk a little more about separation from life.

JOKO: Well, the minute there is a disagreement between ourselves and another person—and we think we're right—we have separated ourselves. We're over here and that nasty person is over there and he is "wrong." When we think this way, we don't have any interest in that person's welfare. What we're interested in is our welfare. So the seamless unity has been broken. For most of us, years of relentless practice are required before we abandon such thinking.

STUDENT: I see that upsets have to do with me not wanting to face what's going on. But I guess I'm still not clear about why the upset is the separation from life.

JOKO: It isn't separation *if* it is nonverbally experienced. But most of the time we refuse to do that. What is it we prefer to do? We prefer to think about our misery. "Why doesn't he see it my way? Why is he so stupid?" Such thoughts are the separating factor.

STUDENT: Thoughts? Not the avoidance?

JOKO: The thoughts *are* the avoidance. We wouldn't think if we weren't trying to avoid the experience of fear.

STUDENT: You mean the thoughts cause the separation?

JOKO: Not if we are fully aware of the thoughts and know they are just thoughts. It's when we believe them that the separation occurs. ("A tenth of an inch of difference and heaven and earth are set apart.") There's nothing wrong with the thoughts themselves, except when we don't see their unreality.

STUDENT: Can we react without there being any thoughts?

JOKO: If we react, thoughts are occurring. They may not be obvious to us, but they are there. For instance, if you insult me I won't react unless I have some thoughts about the insult. But when we begin to judge people as right and wrong we've separated ourselves; right and wrong are just thoughts and not the truth.

STUDENT: What you're describing can sound like being very passive or being a doormat. Could you address that?

JOKO: No, it's not about being passive at all. We can't deal intelligently with the issues of life if we are stuck in our thoughts about them. We have to have a view that's larger than that. Zen practice is about action, but we can't take adequate action if we believe our thoughts about a situation. We have to *see directly* what a situation is. It's always different than our thoughts about it. Can we take intelligent action without really seeing—not what we want to see or what would suit our comfort, but just what's there? No, I'm definitely not talking about being passive and not taking action.

STUDENT: When I see people who are centered in what's happening, they act much faster and better than I do. I noticed in the Mother Teresa film that she went right into a disaster area and started to work.

JOKO: Just doing. Just doing. She didn't stop and ponder, "Should I do this?" She saw what had to be done and did it.

STUDENT: It seems a lot to expect of ourselves just to be on the razor's edge, because our memories enter into each moment, what's happened in our life.

JOKO: Memories are thoughts and nearly always selective and biased. We may completely forget the nice things our friend has done if there is one incident that we see as threatening. Practice does expect a great deal of us; but we are just living this moment; we don't have to live 150,000 moments at once. We are only living one. That's why I say, "What else do you have to do?—you might as well practice with each moment as not."

STUDENT: Well, I think the razor's edge is sort of a boring place. We usually take notice when we have a huge emotional outburst, but when we do the dishes, there's not much to say. It's just . . .

JOKO: Right. If we could just do what there is to be done in every second, there could be no problem; we would be walking the razor's edge. But when we feel upset, then the razor's edge seems alien to us because to experience upset is to experience unpleasant bodily sensations. Because they are unpleasant, we can't see the upset as basically the same life as doing the dishes. Both are utter simplicity.

STUDENT: If we give up our belief in our thoughts, it seems scary— how would we know what to do?

JOKO: We always know what to do if we are in touch with life as it is.

STUDENT: For me the razor's edge is the experiencing of what the moment is. As I continue practice I find more and more that the simple mundane things of life aren't as boring to me as they once were. There is sometimes a depth and beauty that I was never aware of.

JOKO: That's so. Once in a while a student comes in to talk with me, and she is sitting well but she complains, "It's so boring! I'm just sitting and nothing's going on. Just hearing the traffic . . ." But just hearing the traffic is the perfection! The student is asking, "You mean that's all there is?" Yes, that is all there is. And none of us wants life to be "just that" because then life is not centered on us. It's just as it is; there is no drama and we like drama. We prefer to "win" in an argument, but if we can't win, we'd rather lose than not have a drama centered on us. Suzuki Roshi once said, "Don't be so sure you want to be enlightened. From where you're looking, it would be awfully dull." Just doing what you're doing. No drama.

STUDENT: Isn't following the breath being on the razor's edge?

JOKO: Indeed it is. I would probably prefer to say "experiencing the body and the breath." And I want to add that, in following the breath, it is best not to try to control it (control is dualistic, me controlling something separate from myself), but just to experience the breath as it is: if it is tight, experience tightness; if it is rapid, experience that; if it is high in the chest, experience that. When the experiencing is steady, the breath will gradually become slow, long, and deep. If attachment to thoughts has markedly diminished, the body and breath will eventually relax and the breath will smooth out.

STUDENT: Why is it a bigger upset when the upset is with someone close to me?

JOKO: Because it's more threatening. If the person who is selling me a pair of shoes announces, "I'm leaving you," I don't care, that's all right with me; I'll get somebody else to sell me a pair of shoes. But if my husband says, "I'm leaving you," it's not in the same ballpark at all.

STUDENT: Is that threat immediate or does it come from a reservoir of unresolved psychological material?

JOKO: Yes, there is a reservoir; but that reservoir is always held in our bodies as contraction in the present moment. When we experience the contraction or tension, we pull up our entire past. Where is our past? It's right here. There is no past, apart from right now. The past is who we are at this moment. So when we experience that we take care of the past. We don't have to know all about it.

But how does the razor's edge relate to enlightenment? Anybody?

STUDENT: It *is* enlightenment.

JOKO: Yes it is. And none of us can walk it all the time, but our ability to do it vastly increases with years of practice. If it doesn't then we're not really practicing.

Let's finish. But please maintain your awareness as much as you can in every moment of your life. And keep the question with you: right now, am I walking the razor's edge?

New Jersey Does Not Exist

We assume that reality is as we see it and that it is fixed and unchanging. For example: if we look outside and see bushes, trees, and cars, we assume that we are seeing things as they are. But that's only how we see reality at ground level. If we are in an airplane at thirty-five thousand feet on a clear day, and we look down, we don't even see people and cars. At that height our reality does not include people, though it will include mountaintops, plains, and bodies of water. As the plane descends our experience of reality changes. Not until the plane is almost on the earth do we have a human landscape, including cars, people, and houses. For an ant crawling along the sidewalk, human beings don't even exist; they're too enormous for an ant. An ant's reality is probably the hills and valleys of the sidewalk. The foot that steps on an ant—what is that?

The reality that we live in needs us to function in certain ways. To do this, we must be distinct from things around us—from the rug, or from another person. But a powerful microscope would reveal that the reality we encounter is not truly separate from us. At a deeper level we are just atoms and atomic particles moving at enormous speed. There is no separation between us and the rug or another person: we are all just one enormous energy field.

Recently my daughter showed me some pictures taken of white blood cells in rabbit arteries. These cells are scavengers whose function is to clean up the debris and unwanted material in the body. Inside the artery you can see the little creatures crawling along, cleaning by forming pseudopodia that extend toward their targets. Reality for a white blood cell is not the reality we see. What is reality to such a cell? We can only observe its work, which is to clean. And right now as we sit here there are millions of these cells inside us, cleaning our arteries as best they can. Looking at the successive shots, one can see the work the cell is trying to do: the cell knows its purpose.

We humans, with probably the most immense gifts of any creature, are the only beings on earth that say, "I don't know the meaning of my life. I don't know what I'm here for." No other creature—certainly not the white blood cell—is confused like that. The white blood cell works tirelessly for us; it's inside of us, cleaning as long as it lives. And of course that's just one of a hundred thousand functions that take place within this enormous intelligence that we are. But because we have a large brain (which is given to us so we can function) we manage to misuse our native gifts and to do mischief that has nothing to do with the welfare of life. Having the gift of thinking, we misuse it and go astray. We expel ourselves from the Garden of Eden. We think not in terms of work that needs to be done for life, but in terms of how we can serve our separate self—an enterprise that never occurs to a white blood cell. In a short time its life will be over; it will be replaced by others. It doesn't think; it just does its work.

As we do zazen and more and more perceive the illusory nature of our false thinking, the state of natural functioning begins to strengthen. That state is always there; but it's so covered in most of us that we simply don't know what it is. We are so caught in our excitement, our depression, our hopes, and our fears that we cannot see that our function is not to live forever, but to live this moment. We try in vain to protect ourselves with our worried thinking: we plot how we can make it nicer for ourselves, how we can be more secure, how we can perpetuate forever our separate self. Our body has its own wisdom; it's the misuse of our brain that screws up our lives.

A while ago I broke my wrist and wore a cast for three months. When the cast was removed I was touched by what I saw. My hand was just skin and bones, very feeble and trembling; it was too weak to do anything. But when I got home from the hospital and started to do a task with my good hand, this little nothing of skin and bones tried to help. It knew what it was supposed to do. It was almost pathetic: this little skeleton, with no power, still wanted to help. It knew its function. As I looked at it, it seemed to have nothing to do with me; the hand seemed to have its own life; it wanted to get in there and do its work. It was moving to see this little scarecrow trying to do the work of a real hand.

If we don't confuse ourselves we also know what we should be doing in life. But we do confuse ourselves. We engage in odd relationships that have no fruits in them; we get obsessed with a person, or with a movement, or with a philosophy. We do anything except live our life in a functional way. But with practice we begin to see through our confusion, and can discern what we need to do—just as my left hand, even when it couldn't function, still made an effort to contribute, to do the work that needed to be done.

When something really annoys us, irritates us, troubles us, we start to think. We worry, we drag up everything we can think of, and we think and we think and we think—because that's what we believe solves life's problems. In fact what solves life's problems is simply to experience the difficulty that's going on, and then to act out of that. Suppose my child has screamed at me and told me I'm a terrible mother. What do I do? I could justify myself to her, explain all the wonderful things I did for her. But what heals that situation, really? Simply experiencing the pain of what's happened, seeing all my thoughts about it. When I do that sincerely and patiently, I can begin to sense my child differently, and I can begin to see what to do. My action emerges from my experience. But we don't do that with the problems of life; instead we spin with them, we try to analyze them or try to find who's to blame for them. And when we have done all that, we try to figure out an action. That's backwards. We've cut ourselves off from the problem; with all our thinking, reacting, analyzing, we can't solve it. The blockage of our emotion thought makes the problem unsolvable.

Once when I was flying across the United States and I knew we were roughly in the middle of the country, I looked down and thought, where is Kansas? There was no way of saying where Kansas was. Yet we really think that there's Kansas, and Illinois, and New Jersey, and New York, when in reality there is just land going on and on. We do the same thing with ourselves. I think I'm

New Jersey and he's New York. I think that New York is to blame for New Jersey's problems. (It sends all its commuters over to New Jersey.) New Jersey, if it thinks of itself as New Jersey, immediately acquires its own set of problems. It has to identify with all the wonderful things about New Jersey; and certainly it hasn't got much use for Pennsylvania over there. In fact these boundaries are arbitrary; but if we indulge in our emotion thought which separates, we think there is a boundary between ourselves and others. If we work with emotion-thought intelligently the boundaries gradually dissolve, and we realize the unity that is always right there. If our mind is open, just dealing with the sensory input that life presents, we don't have to strive for something called "great enlightenment." If New Jersey does not have to exist as a separate entity, it doesn't have to defend itself. If we do not have to exist as a separate entity, we have no problem. But our lives are absorbed with the question of "What would be best for *me*? How can I make things nice for *me*?" And we include other people and things only to the degree that they are willing to play our game. Of course they're never *really* willing because they're all doing it too. So the game never can work. For example, how can a marriage work if two people see themselves as New Jersey and New York? It may look as if it works sometimes; but until they see that there is no boundary (and that means the dissolution of the blockage of emotion-thought), there will be a running war between them.

We haven't learned how to live as human beings; we've created a false world pasted on top of the real one. We confuse the map with reality itself. Maps are useful; but if we just look at the map we don't see the unity that is the United States. There is no Kansas as a separate entity. Like the white blood cell, we're designed to have a certain function within this enormous energy pattern that we are. We do have to have a certain form in order to function, just as the white blood cell has to form its little legs to do the cleaning. We have to have a certain form in order to function; we have to look as though we're separate, in order to play this wonderful game we're in. But the trouble is we're not playing the real game. We're playing a game that we pasted on top of the real one; and

that game will wreck us. If we don't see through it, we live out our
days on this earth without ever enjoying any of it. The game when
well played is for the most part a good game. It includes sorrow
and joy and disappointment and problems. But it's always real
and rich, and it's not unsatisfying or without meaning. The white
blood cell does not ask "What is the meaning of life?" It knows.
And when we break through that blockage of emotion thought,
then we also begin to know who we are and what we are meant
to do in life. What are we to do in life? If we don't confuse ourselves too much with our false thinking, we know. When we turn
away from our personal obsession with ourselves, the answer
becomes obvious. But we don't do that easily, because we're
attached to our self-righteous thinking.

But occasionally, when we practice meticulously, there are
moments (sometimes hours, sometimes days) when, though we
still have the same problems, it's OK. The longer and harder we
have been practicing, the longer this state lasts. That is the enlightened state: when we simply see, "Oh, that needs to be done
next—OK. I must go to the dentist on Tuesday; I may not like it,
but that's fine. I have to spend two hours with that tiresome person . . . well, I'll just see how it goes." It's amazing: the flow is so
easy. And then (if we're not careful) our confusion begins to take
over again. And that clarity and power begin to fade. The mark of
years of good practice is that the periods of clarity get longer and
the confusion gets less.

Of course, no matter how long we have been practicing, there
are parts of life that seem muddled and confused. "I don't know
quite what's going on here." But to be willing to be muddled and
confused is, paradoxically, the clarity itself. Over and over I hear
from my students, "Right now I feel confused about my practice,
a little nervous. I don't seem to be able to get things clear." What
do we do then? In fact, all of life is like that. For each of us, every
day presents periods like that. What do we do? Instead of trying
to figure out the confusion and nervousness so we can arrive
somewhere else, we ask ourselves, "What does confusion feel
like?"—and we settle back into the body and its sensations, keep-

ing track of the thoughts that float around. Before we know it we're back on track.

In times of confusion and depression the worst thing we can do is to try to be some other way. The gateless gate is always right here when we experience ourselves as we are, not the way we think we should be. When we truly do that, the gate opens — though it opens when it should open, which is not necessarily when we want it to open. For some people an early opening would be a disaster. I'm skeptical of pushy practice; to push for clarity too fast just creates more problems. Of course the alternative is not to sit and do nothing. We must maintain awareness of bodily sensations, thoughts and whatever else is here, whatever it is. We don't need to judge our sittings as good or bad. There's just, "I'm here, and I'm at least aware of some of my life." And as I sit meticulously, that percentage tends to go up.

A part of us is like the white blood cell: it's always there and it knows what to do. It wants to function. Practice is not a mystical push into some other place, God knows where. The absolute isn't somewhere else; where else could it be but right here? My nervousness is what? Since it exists right here now, if I'm nervous, that is nirvana, that is the absolute. That's it. There's no place to go; we're always right here. Where could we be except where we are? We always are as we are. Our innate intelligence knows who we are, it knows what we're about in this world if we don't muck it up.

Religion

People who come to Zen centers are often upset of disillusioned with their past experiences of religion. The original meaning of the word "religion" is interesting: it's from the Latin *religare*, which means "to bind back, to bind man and the gods." *Re* means "back," and *ligare* means "to bind."

What are we binding? First of all, we bind our self to itself — because even within ourselves we're separated. And we bind ourselves to others, and eventually to all things, sentient and insentient. And we bind others to others. Anything that is not bound together is our responsibility. But most of the time our task is to bind ourselves to our roommate, to our work, to our partner, to our child or friend, and then to bind ourselves to Sri Lanka, to Mexico, to all things in this world, and this universe.

Now that sounds nice! But in fact, we don't very often see life that way. And any true religious practice is to see once again that which is already so: to see the fundamental unity of all things, to see our true face. It's to remove the barrier between ourselves and another person or another thing: to remove or to see through the nature of the barrier.

People often aks me, if this fundamental unity is the true state of affairs, why is it almost never seen? It's not because of a lack of the right scientific information; I've known a lot of physicists who had the intellectual knowledge, yet their dealings with life did not reflect this awareness.

The main cause for the barrier, and the main reason we fail to see that which is already so, is our fear of being hurt by that which seems separate from us. Needless to say, our physical being does need to be protected or it can't function. For instance, if we're having a picnic on a train track and a train's coming, it's quite a good idea to move. It's necessary to avoid and to repair physical damage. But there's immense confusion between that kind of hurt, and other less tangible occurrences that seem to hurt us. "My lover left me, it hurts to be alone." "I'll never get a job." "Other people are so mean." We view all these as sources of hurt. We often feel we have been hurt by other people.

If we look back on our lives we can make a list of people or events who have hurt us. We all have our list. Out of that long list of hurts we develop a conditioned way of looking at life: we learn patterns of avoidance; we have judgments and opinions about anything and anyone that we fear might hurt us.

Our innate capacities are exerted in the direction of avoidance, in the direction of complaining, of being the victim, of trying to set things up so that we maintain control. And the true life, the fundamental unity, escapes us. Sadly enough, some of us die without ever having lived, because we're so obsessed with trying to avoid being hurt. One thing we're sure of: if we have been hurt, we don't want it to happen again. And our mechanisms for avoidance are almost endless.

Now in many religious traditions, and particularly in the Zen tradition, there is great stock placed in having what are called "openings" or enlightenment experiences. Such experiences are quite varied. But if they are genuine they illuminate or bring to our attention that which is always so: the true nature of life, the fundamental unity. What I have found, however (and I know many of you have found it too), is that by themselves they're not enough. They can be useful; but if we get hung up on them they're a barrier. For some people these experiences are not that hard to come by. We vary in that respect—and the variation is not a matter of virtue, either. But without the severe labor of unifying one's life these experiences do not make much difference. What really counts is the practice that we have to go through moment by moment with that which seems to hurt us, or threaten us, or displease us—whether it's difficulty with our coworkers, or our family, or our partners, anyone. Unless in our practice we've reached the point in our practice where we react very little, an enlightenment experience is largely useless.

If we truly want to see the fundamental unity, not just once in a while, but most of the time—which is what the religious life is—then our primary practice has to be with what Menzan Zenji (a Sōtō Zen scholar and teacher) calls the "barrier of emotion-thought." He means that when something seems to threaten us, we react. The minute we react a barrier has come up and our vision is clouded. Since most of us react about every five minutes, it's obvious that most of the time life is clouded over for us. We are caught within our own selves, we're caught in this barrier.

Our primary practice is with this barrier. Without such practice, without understanding all the ins and outs of the barriers we erect—which is not an easy matter—we remain enslaved and separate. We may see our true face once in a while, but still we find it impossible to be ourselves, moment by moment. In other words the religious life has not been realized: humanity and the gods remain separate. There's me and there's the life out there that I view as threatening—and we do not come together.

This barrier of emotion-thought often takes the form of a vacillation between two poles. The one pole is conformity: sacrificing to the gods, sacrificing ourselves, pleasing life, pleasing others, being good, trying to be an ideal person, stifling what is true for us at any given moment. This is the person who tries to be good, who tries to practice hard, who tries to be enlightened, who tries, tries, tries. Such efforts are extremely common, particularly among Zen students. But if we practice with intelligence we begin to sense the conforming style we've been immersed in and then we tend to swing to the opposite pole, to another kind of slavery: rebellion or nonconformity. People then insist, "No one can tell me what to do! I need my own space, and I want everyone to stay out of it!" In this phase we judge others harshly, and have strong negative opinions. Instead of seeing ourselves as inferior and dependent, we now see ourselves as superior and independent. These states (conformity and nonconformity) flow into each other moment by moment. In the first years of practice most people swing out of the first stage and into the second. At that point it may seem that their lives are getting worse, not better! "Where's that nice person I used to know?" Both states are slavery, however; we are still reacting to life. Either we conform to it or we rebel against it. People and the gods are still separate.

We all swing between these two stages. One day last week I made up my mind at 9 A.M. that I was going to answer a letter, a difficult letter I didn't want to answer. At 3 P.M. it dawned on me that I still hadn't answered that letter. I'd found fifteen things to do from nine to three that were not answering the letter. My initial response was, "I *should* answer the letter." That's conformity. "It's required of me, I should do it." The second is, "You can't make me. I don't have to do that. I can leave it sitting there." But the minute the observer sees both states, what happens? When I observed both thoughts I sat down and answered the letter.

What is the resolution? What resolves that continual battle within ourselves? What brings us and the gods together? Until we understand the riddle we're caught in it. The first thing to see is what we're doing. And when we sit that will reveal itself. At first we'll have the thought, "I should do that." And if we sit a while longer the second thought will come up: "I don't want to do that." We begin to see that we swing between these thoughts, back and forth, back and forth.

In this whole back-and-forth process, there's nothing but separation. How do we resolve it? We resolve it by experiencing that which we don't want to experience. We need to experience nonverbally the uncomfortableness, the anger, the fear that is sitting beneath this vacillation between the two poles. That's true zazen, true prayer, true religious practice. Eventually the anger (as physical experience) will begin to shift. If we're *really* upset, the shifting may take weeks or months. But if we surrender to the experiencing, if we "embrace the tiger," it will always shift—because when we are the experience itself, there is no subject and no object. And when there's no subject or object, the barrier of emotion-thought drops and for the first time we can clearly see. When we can see, we know what to do. And what we do will be loving and compassionate. The religious life can be lived.

As long as we don't feel open and loving, our practice is right there waiting for us. And since most of the time we don't feel open and loving, most of the time we should be practicing meticulously. That's the religious life; that's "religion"—though we don't have to use these words. It's the reconciliation of people and their separate notions, the reconciliation of our viewpoints of how it should be, of how people should be, the reconciliation of our fears. The reconciliation of all that is the experience—of what? Of God? Of just what is? The religious life is a process of reconciliation, second by second by second.

And each time we go through this barrier something changes within us. Over time we become less separate. And it's not easy, because we want to cling to that which is familiar: being separate, being superior or inferior, being "someone" in relationship to the world. One of the marks of serious practice is to be alert and to recognize when that separation is occurring. The minute we have even a passing thought of judging another person, the red light of practice should go on.

We all do some harmful actions that we are not conscious of doing. But the more we practice, the more we see that which we previously could not see. That doesn't mean we'll ever see it all — there's always something we do not see. And that's not good or bad; that's just the nature of things.

So practice is not just coming to sesshin or sitting each morning. That's important, but it's not enough. The strength of our practice, and the ability to communicate our practice to others, lies in being ourselves. We don't have to try to teach others. We don't have to say a thing. If our practice is strong it shows all the time. We don't have to talk about the dharma; the dharma is simply what we are.

Enlightenment

Someone said to me a few days ago, "You know, you never talk about enlightenment. Could you say something about it?" The problem with talking about "enlightenment" is that our talk tends to create a picture of what it is — yet enlightenment is not a picture, but the shattering of all our pictures. And a shattered life isn't what we are hoping for!

What does it mean to shatter our usual way of seeing our life? My ordinary experience of life is centered around myself. After all, *I* am experiencing these ongoing impressions — I can't have *your* experience of *your* life, I always have my own. And what inevitably happens is that I come to believe that there is an "I" central to my life, since the experiences of my life seem to be centered

around "I." "I" see, "I" hear, "I" feel, "I" think, "I" have this opinion. We rarely question this "I." Now in the enlightened state there
is no "I"; there is simply life itself, a pulsation of timeless energy
whose very nature includes—or is—everything.

The process of practice is to begin to see why we do not realize
our true nature: it is always our exclusive identification with our
own mind and body, the "I." To realize our natural state of enlightenment we must see this error and shatter it. The path of practice
is deliberately to go against the ordinary self-absorbed way of life.

The first stage of practice is to see that my life *is* totally centered
around myself: "Yes, I do have these self-centered opinions, I do
have these self-centered thoughts, I do have these self-centered
emotions. . . . I, I, I, I, I have all these from morning until night."
Just this awareness is in itself a great step.

Then the next stage (and these stages may take years) is to
observe what we do with all these thoughts, fantasies, and emotions, which usually is to cling to them, to cherish them, to believe
that we would be miserable and lost without them. "Without that
person I will be lost; unless this situation goes my way I can't
make it." If we require that life be a certain way, inevitably we
suffer—since life is always *the way it is,* and not always fair, not
always pleasant. Life is not particularly the way we want it to be,
it is just the way it is. And that need not prevent our enjoyment
of it, our appreciation, our gratitude.

We are like baby birds sitting in their nest waiting for mommy
and daddy to put food in their beaks. That's appropriate for baby
birds, though mommy and daddy have more freedom, flying
around all day. We may think we don't envy the life of baby birds;
but we do just what they do, expecting life to put its little goodies
in our mouths: "I want it my way; I want what I want; I want my
girlfriend to be different; I want my mother to suit me; I want to
live where I want to live; I want money . . . or success . . . or, or,
or . . ." We are like the baby birds except that we hide our greed,
they don't.

In a nature documentary film, a mama bear is shown bringing
up her little ones. She teaches her cubs to hunt, to fish, to climb,

to do what cubs need to learn for survival. Then one day she chases all of them up a tree. And then what does she do? Mama bear just *leaves*, and doesn't even look back! How do the cubs feel about this? They probably feel terrified, but the path of freedom *is* to be terrified.

We are all baby birds or baby bears, and we would like to find some piece of mama-life to hang on to—preferably in eighteen different ways, but at least in *one*. None of us wishes to hop out of our nest, because that's terrifying. But the process of becoming fully independent (or of experiencing that we already are that) is to be terror, over and over and over. We fight against being free, against the abandonment of our dream that eventually life will be exactly as we wish it, that it will shelter us; that's why practice can seem difficult. Zazen is to free us to live a soaring life which, in its freedom, its nonattachment, is the enlightened state—just being life itself.

In our first years of practice we do zazen to understand our attachment process in its gross aspects; and then over the years we practice with our more subtle (and even more poisonous) attachments. Practice is for a lifetime. There is no end to it. But if we truly practice we definitely realize our own freedom. A cub who has been away from mama for two or three months may not have the strength and skill of mama, but still it is doing well and probably enjoying life more than the little bear who has to trail mama everywhere she goes.

Daily zazen is essential; but because we are so stubborn we usually need the pressure of long sitting to see our attachments. To sit a long sesshin is a major blow to our hopes and dreams, the barriers to enlightenment. And to say that there is no hope is not at all pessimistic. There can be no hope because there is nothing but this very moment. When we hope, we are anxious because we get lost between where we are and where we hope to be. No hope (nonattachment, the enlightened state) is a life of settledness, of equanimity, of genuine thought and emotion. It is the fruit of true practice, always beneficial to oneself and to others, and worth the endless devotion and practice it entails.

VIII. CHOICES

VILE CHOICES

From Problems to Decisions

Sometimes people who come to the Center—especially new people—say that what they really want here is to find a spiritual life, or a life of oneness, a life where they feel that they are one with everything instead of separate. There's nothing wrong with that; that's what we *are* doing here.

Still, I don't think any of us could say what "spiritual life" *is*. So we talk mostly about what it *isn't*. There's a famous passage in Zen literature: "A tenth of an inch of difference, and heaven and earth are set apart." What is that all about? What is this tenth of an inch of difference when "heaven and earth are set apart," when the wholeness of life is disrupted (or we think it is)? From the absolute point of view, nothing could disrupt it, but from our relative point of view, something doesn't feel right. The essential wholeness of life seems to be unavailable to us. Sometimes we have glimpses of it—but most of the time we don't.

For example: at Chirstmas time people are either enjoying themselves or going crazy. We can combine them both sometimes! It is a season when we tend to be aware of our anxiety, our disruption. But also, as we approach the New Year, we sense that the holiday season is a time of *turning*, and no human being can take that turning lightly. We have only a certain number of turning years on the planet. For people who are at all sensitive, the turning of the New Year is crucial. We need to see this tenth of an inch of difference—to see what it is, how it is related to the turning in our life.

A passage from the Bible says, "As a man thinketh in his heart, so is he." This unease we are speaking of, this separation, this tenth of an inch of difference comes from how a man "thinketh in his heart." (The word "heart" here doesn't refer to some emotional characteristic, but rather the heart of the matter, the truth of the problem, the central core, as in the Heart Sutra.) And "as a man thinketh in his heart"—as a man begins to see the truth of his life—then he is that. Now, the more we see what the truth of our life is, the more we see what the tenth of an inch of difference is. And that leads me to two words that sound similar and are often used interchangeably: decisions and problems.

Life from morning to night is nothing but decisions. The minute we open our eyes in the morning we make decisions: Should I get up now or should I get up five minutes later? Particularly, should I get up and sit! Should I have a cup of coffee first? What should I have for breakfast? What should I do first today? If it's a day off, should I go to the bank? Or should I just have a good time? Should I write the letters that I haven't written? From morning to night we make one decision after another and that's normal; there's nothing strange about it. But we see life in terms of problems, not decisions.

For instance, you may say, "Well, it's one thing to decide whether to go to the bank or the supermarket first; that's just a simple decision. But what I've got in my life is a problem." It might have something to do with our job; perhaps we have a job we absolutely hate. Or maybe we've lost our job . . . or whatever. We don't think that is just a decision, we think it's about a problem. We all worry about how to solve the problems of our life; we all see life as a problem, at least some of the time. Another example: "I'm working in San Diego, I have a nice girlfriend here and I like the climate—but gosh, I've got a great offer in Kansas City, with more money." We feel we can't just make a decision—and so we have a problem. And this is where human life can snarl itself up, and the tenth of an inch of difference begins to appear.

What should we do about our problems—as opposed to stewing, analyzing, muddling, feeling we just don't know what to do? I'm not talking about the little issues; we'll make some decision and move from there. But when we have a major issue in our life—"Should I enter this relationship?" "Should I finish this relationship?" "If I want to end a relationship, how do I do that?"—we are baffled about what to do. And that's where the quotation applies: "As a man thinketh in his heart, so is he." What really decides any problem is the way we think in our hearts. How we see what our life is. Out of that we make our decision.

Suppose we've been sitting for two years; we may not know it, but we would probably look differently at the problem of how to end a relationship than we did before we began to practice, because who we think we are and who we think he or she is have altered. Serious practice changes the way we see our life, and so what we do with that life begins to shift. People want a mechanism for making decisions, for solving problems. There can be no fixed mechanism. But if we know more and more who we are, out of that we will make our decision.

For instance: suppose we said to Mother Teresa, "Well, Mother Teresa, you might consider living in San Francisco rather than Calcutta; the night life is better. You would have nicer places to go out to dinner. The climate's easier." But how does she make her decision? How did she make her decision to stay in the hellish part of Calcutta where she works? Where did that decision come from? "As a man thinketh in his heart, so is he." She would probably call it prayer: from her years of being with herself, she sees where she works or what she does as not a problem, but just a decision.

The more we know who we are, the more our problems shift into, "I am this therefore I will do that, or I am to some degree willing to do that." And we will sometimes choose things that look to other people very trying, very unpleasant. "What do you mean, you're going to do that? I wouldn't do that." But if for me in my heart that's what I feel I am and the way my life wants to express itself, then there is no problem.

So when something in our life looks insoluable, it means that we think that there's a problem out there that we look at as an object, like a grapefruit. We have not viewed that problem as ourself. And one way to get a problem to change into a decision is to sit with it, doing zazen. For instance, that decision about where I'm going to work: if I sit with it the thoughts will come floating up about all my reservations, or my this or that about working at an out-of-state job. I keep labeling them, letting them float through; I worry, analyze, or fuss. And I keep returning to the direct experience in my body of the truth of this matter. I just sit with the tension and contraction, breathing through it. And in doing this I get more in touch with who I am and the decision begins to be clear. If I feel completely muddled it isn't that there's a problem that I have to find some way to solve; I just don't know who I am in connection with that problem.

For example, suppose I don't know whether I should marry one man for his money or another man for no reason—I just like him. If that question can even occur to me, there's something about myself I don't know. The problem isn't out there. The problem is here: I don't know who I am. When I know who I am, like Mother Teresa I will have no problem in knowing what to do. And as I know more and more who I am, I begin to strip my life down to what my life truly needs. I no longer find suddenly that I absolutely *have* to have that, or that, or that. It's not that I give them up, it's just that I don't really need them so much. Most people who sit for years find their lives considerably simplified—not because of some virtue, but because, needing less, desires naturally drift away. People who know me now can't believe it, but years ago I never could go to work without nail polish and lipstick (all matching): I was uncomfortable if they weren't just so. And while I never had a lot of money, I always had to have nice clothes. Now there's nothing wrong with looking nice; I'm not saying that. I'm saying that when self-centered desires are a central concern, then you will have trouble with your decisions. They will seem like problems. But as you practice, as your central concern about what you really want for your life changes, desires and indecision drift away.

So we have difficulty at Christmas, endlessly dashing around fulfilling everybody else's desires. We have to know for ourselves what is central. Then we know how much is appropriate to do. Of course this knowledge of who we are is always fragmentary, incomplete, even elementary. Nevertheless as we practice, more and more, life isn't about problems or complaints.

I don't want to imply that we never should have fun. We will have a desire for just about as much fun as suits our picture of who we are right now. If we need a lot of escape time that's just the way we see ourselves and our life. But over time that will diminish. Because we can't get in touch with this central core, the heart of ourselves, without everything shifting and changing around it. T. S. Eliot wrote of that still point on which the universe turns. That still point isn't a something. As we practice, more and more we know what it is. But without persistent, patient practice, which for most of us is zazen, we tend to be confused. For example, we may demand of ourselves a lot of self-sacrifice. Sometimes sacrificing ourselves for another can be very bad for that other person. Sometimes it's the thing to do. When we face a decision whether or not to do something for another, and finally say, "No, I won't do that for you"—where does that come from, the ability to make a wise decision? It comes from increasing clarity about who we are and what our life is about. Over the years I do less and less for people, at least in the sense that I used to. Whenever anyone with a little difficulty knocked on the door, I used to feel that I had to talk to them right then. I put myself first a lot of the time now. And that's not necessarily being selfish, it might be the best thing to do.

The knowledge of what needs to be done slowly clarifies with practice. And decisions become just decisions, not heart-rending problems. A sesshin is a way of pushing us past that part of ourselves that wants to jitter about our problems. By its very structures it gives us, almost whether we want it or not, a space in which we see more clearly. Still the most important thing is daily sitting. I'm not talking about sitting just any old way. If that sitting isn't intelligent, it's almost worse to do it than not to do it. We have to know what we're doing. Otherwise, we build a new fantasy world, which is probably more harmful than not sitting at all. OK, questions.

STUDENT: It seems that if you have ideas about what's right and wrong, it interferes.

JOKO: Oh sure it does! Because those are thoughts, and thoughts in my head about what is right and wrong are my personal viewpoint—and usually emotionally based—which interferes with my seeing myself or anyone else clearly.

STUDENT: I would think the answer would be to see reality just the way it is.

JOKO: That's fine. Again, what that means in terms of actual practice may not be so simple. "A tenth of an inch of difference" . . . what is it?

STUDENT: If there's something that I have planned to do and then something else turns up so there are two potential scenarios I must choose from—in that gap I start to get upset and have self-centered thoughts . . .

JOKO: You have a "problem," right?

STUDENT: And more than a tenth of an inch!

JOKO: *More* than a tenth of an inch! All right!

STUDENT: The difference perhaps has to do with recognizing what is unto me, the responsibilities that are unto me.

JOKO: Do you always know what they are?

STUDENT: No!

JOKO: Well, what creates that tenth of an inch of difference, so that we can't see? We all have responsibilities and obligations, but we confuse them too, and turn them into problems. What is it we do that creates that tenth of an inch of difference?

STUDENT: We want things.

JOKO: We want things. Yes.

STUDENT: We have thoughts about giving things.

JOKO: And we can only truly give when we don't need some return. Right? I want. I want. I want. I want. Just to recognize that I want, I want, I want my life to be the way I want it to be, and not any other way—that has a lot to do with that tenth of an inch of difference. And we all want life to be according to our pictures of it, preferably comfortable. Pleasant. What else? Full of hope for the future? There is no future! "It's all gonna be nice someday." Who knows?

STUDENT: For me, it's surrender. If I can surrender to what's happening, then I don't bring up all this stuff that I bump into.

JOKO: If we can truly surrender, that's fine. But what stands in the way of surrender? Me. And what does that me consist of?

STUDENT: Anger. I want it another way! It's not the way I planned it.

JOKO: Right, and these are thoughts. If we saw those thoughts as being just thoughts we could return to what needs to be done.

STUDENT: When you see a problem, should you use your will to change it?

JOKO: You're talking about the difference between decisions and problems. If you really view any problem as yourself, instead of looking at it as a problem to be solved, you can ask, "What's going on here?" What you see going on is usually your own anger, fear, thoughts. And the more you get acquainted with them and the accompanying physical tension, whether or not you should attempt change becomes obvious. I'm not saying never to change something. But it will be obvious, just as it is to Mother Teresa.

STUDENT: Is that the cure?

JOKO: The cure? There is no cure; but the minute you embrace life and make it yourself, you see just what it is, what's going on. Then that tenth of an inch is gone, you understand? Because the problem isn't there. It's just myself. It's not frightening then. As we patiently sit, we tend to see more and more what to do. It's not such a mystery. And we know when to change things and when not to change them. As the saying goes, we gain acceptance of the things we cannot change, courage to change what needs to be changed, and the wisdom to know the difference.

STUDENT: What makes us want to do what is appropriate?

JOKO: We always want to do what is appropriate when we are in touch with ourselves. "As a man thinketh in heart, so is he." Not only "so is he," so *does* he. So he acts.

Turning Point

We all want a life of freedom and compassion, a fully functioning human life. And a fully functioning human life can be attached to *nothing*, not to a practice or a teacher or even the Truth—if we're attached to the Truth, we can't see it.

I saw a story on television news about a man who found boxes and boxes of machine parts. He didn't have a clue what they were for, but he enjoyed putting things together, and the mystery made it more exciting. So he began his labor. It took him ten years to fit together all the thousands of pieces, some large, some small. When he had finished his work, he had created a shiny, new, and beautiful Model-T Ford. But (evidently he didn't have a wife!) he had built it *in his living room*. So, after some soul-searching, he knocked out the front wall of his living room and pushed the Model-T onto the porch—a definite improvement. The porch was elevated by four feet, however, so he had to build a ramp down to the yard. Finally he was able to move the car down into the yard and the street, and at last the Model-T could be a real, functioning car.

This is a wonderful story, because it's like what we do with our lives. We build a strange creation we call "myself." Unfortunately we're not very skilled in building the self and, when we have built it, we have an uneasy sense that our self (like the Model-T) is confined, that the walls are pushing in. The self may look good, even impressive, but still it feels uncomfortably constricted to us.

Now comes the crucial choice: there are two ways to go once we sense the confinement and anxiety of "myself." One way is to pretend that our living space was really meant to hold a Model-T— and we may decorate the walls or create some subterfuge with mirrors so that there is the illusion of ease and space. The other way is to recognize that this fixed, constricted "self" must in some way be moved, so we break through to a place where there is more light and air.

At this point (when we begin to examine the Model-T, this self we have built) our practice really begins. We no longer hope to fix up the surroundings, the environment; instead we move the Model-T, the self, out so we can examine it. That's not the final stage, of course; the final stage of human life isn't to examine and analyze the self, to see how it ticks; the final stage is to get our life out on to the street where it can fully function.

It's the pain of the confining walls that first motivates most of us to move; we know we must do *something* about the walls. And it's a major step just to move that Model-T out on the porch where we can get a little more light on it, a little more space and perspective. In practice, it is the crucial turning point. So what must we do in order to foster a turning point?

Let's consider the idea of "renunciation." We often feel that for our life to have a new start, the old one must be renounced. What might we consider renouncing? We might renounce the material world, as we conceive it; or we might renounce our mental and emotional world.

Many traditions do encourage giving up all material possessions. Monks traditionally have kept just a small box containing a few necessities. Is that renunciation? I'd say no; though, it's useful practice. It is as if we have felt that our evening meal was not complete without dessert; so we go without dessert for a time as a means of learning about ourselves; and that is good practice.

Then we may feel that whatever is going on in our thoughts and emotions is not OK: "I should be able to renounce all that; I should be able to get rid of it. I'm bad for thinking or feeling this." But that's not renunciation either; it's playing with notions of good and bad.

Some of us make one final effort. Because we are confused and discouraged about our daily lives, we finally decide, "I have to go for Realization—I must live a completely spiritual life and renounce everything else." And that's great if we understand what it means. But of all the misinterpretations of renunciation, the most insidious come in this realm of so-called spiritual practice in which we have notions such as "I should be pure, holy, different from others . . . perhaps live in a remote, quiet environment." And *that* has nothing to do with renunciation, either.

So what *is* renunciation? Is there such a thing? Perhaps we can best clarify it by considering another word, "nonattachment." We often think that if we fiddle with the surface events of our lives, trying to alter them, worrying about them or ourselves, we are dealing with the matter of "renunciation"—whereas in fact we do not need to "renounce" anything, we need only to realize that true renunciation is equivalent to nonattachment.

The process of practice is to see through, not to eliminate, anything to which we are attached. We could have great financial wealth and be unattached to it, or we might have nothing and be very attached to having nothing. Usually, if we have seen through the nature of attachment, we will tend to have fewer possessions, but not necessarily. Most practice gets caught in this area of fiddling with our environment or our minds. "My mind should be quiet." Our mind doesn't matter; what matters is nonattachment to the activities of the mind. And our emotions are harmless unless they dominate us (that is, if we are attached to them)—then they create disharmony for everyone. The first problem in practice is to see that we *are* attached. As we do consistent, patient zazen we begin to know that we are nothing *but* attachments: they rule our lives.

But we never lose an attachment by saying it has to go. Only as we gain awareness of its true nature does it quietly and imperceptibly wither away; like a sandcastle with waves rolling over, it just smooths out and finally—where is it? What was it?

The question is not how to get rid of our attachments or to renounce them; it's the intelligence of seeing their true nature, impermanent and passing, empty. We don't have to get rid of anything. The most difficult, the most insidious, are the attachments to what we think are "spiritual" truths. Attachment to what we call "spiritual" is the very activity that hampers a spiritual life. If we are attached to anything we cannot be free or truly loving.

So long as we have any picture of how we're supposed to be or how other people are supposed to be, we are attached; and a truly spiritual life is simply the absence of that. "To study the self is to forget the self," in the words of Dōgen Zenji.

As we continue our zazen today let's be aware of the central issue: the practice of nonattachment. Let us diligently continue, knowing it can be difficult and knowing that difficulty is not the point. Each of us has a choice. What will it be? A life of freedom and compassion—or what?

Shut the Door

In the 1960s Hakuun Yasutani Roshi began making annual visits to teach the dharma in America. During each visit he would conduct a week-long sesshin here in Southern California. Like others who began Zen practice with Yasutani Roshi during these visits, I would practice intensively with him for seven days each year, and for the rest of the year continue doing zazen on my own. Those sesshins were extremely difficult for me, and I'd have to say that if there was ever a muddled practice it was mine. But having the opportunity to study with him, even though it was only for one week each year, and to see what he was—humble, gentle, vital, and spontaneous—was enough to keep me going.

He was quite old when I knew him, in his eighties and having some physical difficulties. When he shuffled into the zendo I wondered if he would be able to make it all the way to his seat. Just a little, bent old man, shuffling in. But when he would begin his dharma talk, I couldn't believe it! It was like a streak of electricity running through the room—the vitality, spontaneity, the total devotion. It didn't matter what he said, or that he used an interpreter. His very presence revealed the dharma: not to be forgotten if it had once been encountered.

Two qualities of Yasutani Roshi struck me most deeply. I would say he was luminous and ordinary at the same time. Looking into his eyes in a formal interview was like looking for ten thousand miles—there was nothing there. It was amazing. Yet, somehow, in that open space there was total healing.

Outside of the zendo he was just an ordinary little man running around with his broom and with his pants rolled up, eating carrots. He loved carrots.

Yasutani Roshi gave me my first experience of what a true Zen master is, and it was a very humbling experience because he was so humble. Radiating from him were freedom, spontaneity, and compassion, the jewel that we all seek in our own practice. But we must be careful that we don't look for the jewel in the wrong place, outside of ourselves, failing to see that our life itself is the jewel—unpolished perhaps, but already perfect, complete and whole.

When you come right down to it, the dharma is quite simple and always available; but the trouble is that we don't know how to see it. Because we don't, this jewel, this freedom, escapes us.

Freedom is such a sticky thing to talk about. Our usual way of looking at freedom is to see it as a matter of being left alone to go where we want to go and do what we want to do. And we hope that something "out there" will give us freedom so, when we are in an unpleasant and restrictive situation, we leave a door open so we can run out the door to new hope and freedom. All of us without exception do this. Which brings us to another sticky word, commitment.

One important aspect of our practice is to look honestly at this constant process of hopes and fears and all the schemes that are a reflection of our lack of commitment to our lives. To do this requires that we shut the door that we like so much to leave open, and turn around and face ourselves as we are. This is commitment, and without it there is no freedom.

Through practice we wear out the fantasies we have about running out the door to something somewhere else. We put most of our effort into maintaining and protecting the ego structure created out of the ignorant view that "I" exists separately from the rest of life. We have to become aware of this structure and see how it works because—even though it is artificial and not our true nature—unless we understand it, we will continue to act out of fear and arrogance. By arrogance I mean the feeling of being special, of not being ordinary. We can be arrogant about anything: about our accomplishments, about our problems, even about our "humility." Out of fear and arrogance we cling to all kinds of self-centered attitudes and judgments, and so create all kinds of misery for ourselves and others.

Freedom is closely connected with our relationship to pain and suffering. I'd like to draw a distinction between pain and suffering. Pain comes from experiencing life just as it is, with no trimmings. We can even call this direct experiencing joy. But when we try to run away and escape from our experience of pain, we suffer. Because of the fear of pain we all build up an ego structure to shield us, and so we suffer. Freedom is the willingness to risk being vulnerable to life; it is the experience of whatever arises in each moment, painful or pleasant. This requires total commitment to our lives. When we are able to give ourselves totally, with nothing held back and no thought of escaping the experience of the present moment, there is no suffering. When we completely experience our pain, it is joy.

Freedom and commitment are very closely connected. When two people make a commitment to each other in marriage they are, in a sense, shutting the door on their chance to escape the heat and pressure that is part of any relationship. But when accepted as part of their commitment, the heat and pressure make for growth and the relationship blooms. I'm not saying that one should commit oneself to any relationship that comes along— that's crazy. What I mean is that our practice is to commit ourselves to our experience in each moment. Just as the commitment of marriage puts us under heat and pressure, so too does zazen. We might even say that the first thing we must do in zazen is marry ourselves. We shut the door and sit quietly with what is, feeling the heat and pressure.

Often people have the idea when they begin practice that it is going to be nice and comfortable. But Zen practice has phases that are anything but pleasant. By just sitting in this very moment, the secure walls of the ego structure crumble, and this can be confusing and painful. Physically experiencing the confusion and pain rather than avoiding them is the key to freedom. We have to embrace the misery, make it our best friend, and go right through it to freedom.

This jewel of freedom is our life just as it is, but if we don't understand the relationship between pain and freedom, we can cause suffering for ourselves and others. We have to be willing to be on the cutting edge, just being there with whatever comes up in each moment. Pride, greed, arrogance, pain, joy—don't try to manipulate what comes up in zazen. By sitting with as much awareness as we can muster, attachments in time just wither away.

When Yasutani Roshi was eighty-eight, on his last birthday before his death, he wrote, "The hills grow higher." The more clearly we see that there is nothing that needs to be done, the more we see that which needs doing. It's a funny thing; when we really share what we have—our time, our possessions, and most importantly ourselves—our life will go smoothly. There is a story of a well fed by tiny springs that always gave a good supply of water. One day the well was covered over and forgotten until somebody uncovered it years later. Because nobody had drawn water from it, the springs had stopped feeding it and the well had dried up. It's the same with us: we can give of ourselves and open further, or we can hold back and dry up.

Zen practice is shutting the door on a dualistic view of life, and this takes commitment. When you wake up in the morning and don't want to go to the zendo, shut the door on that. Put your foot out of bed and go. If you feel lazy during work, shut the door on that and do your best. In relationships, shut the door on the criticism and unkindness. In zazen, shut the door on dualism and open up to life as it is. Very slowly, as we learn to experience our suffering instead of running from it, life is revealed to us as joy.

Commitment

Once upon a time there was a young man who was madly in love with a beautiful but wicked young lady. The beautiful and wicked young lady wanted him to have no thought in the world of anything except her, and so she announced, "I will only pledge myself to you if you cut off your mother's head and bring it to me."

Now the young man loved his mother. But he was so infatuated with the wicked lady that he could hardly wait to do her bidding. So he hurried home and cut off his mother's head. He grabbed her head by the hair and hurried out into the night, because he could hardly wait to get back to his wicked lover. With his mother's head in his hand, he was racing down the street as fast as he could go when the head spoke to him: "Please don't hurry, my son; you might fall and hurt yourself."

The story is about a mother's undying love and her unalterable commitment. Commitment and true love are twin sisters. The word "commit" is from the Latin *committere*, which means to join, to entrust, to connect. It means to deliver a person or a thing into the charge of another.

To understand commitment we must increasingly intuit the nature of reality—not just in our heads, but in our guts: who we are and what everything is. We may feel we are already committed to a specific job or person; but true commitment is something deeper. Our commitment will lack power and resolve unless we understand our basic vows, which are about a commitment to *all* sentient beings, not just to any particular one. In our usual notions of commitment we tend to think things like, "Well, now that we're committed to each other, obviously you should be a certain way: you should love only me, you should spend most of your time with me, you should put me first in every way . . ." If we are committed to our work we become possessive: it's *our* work, or *our* project, or *our* business, or *our* profits. We may also say, "Because I'm committed, I must be a certain way in this commitment." In our usual notions of commitment the object of commitment becomes in our eyes an object we possess, an investment that should give us safety and happiness.

In fact our commitments are usually a mixture of our Buddha-nature—the part of us that can say, like the mother in the story, "Whatever you do, I love you, I wish the best for you"—and the part of us that says, "I'm committed to you *if . . .*" What a poisonous "if" that is! True commitment and true love have no "ifs." They are not shaken by passing circumstances. As Shakespeare wrote, "Love is not love which alters when it alteration finds."

Commitment cannot be enforced by nagging, by anger, by going on strike, by any of our maneuvers to please, though we try all of these. It cannot be forced in any way. To deepen our commitment we must be the witness of our maneuvers and tricks, the witness of our subtle and not-so-subtle attempts to get what we want, which is always safety and security for ourselves. The mother in the little tale certainly didn't have safety and security; she only had her head. But even in death she wished the best for her son. Of course we're not like that. We're human.

I would never tell anybody, "Just commit yourself to somebody—and then battle it out from there." Even if we've spent months and years deciding that "this is the one," we'd probably just begun to commit ourselves. We fool ourselves and others if we think that because we've made some promises, we have made a commitment.

In commitment we shut the door. Since we're not realized Buddhas, we can't (or won't) commit to just anyone. Still, after much worry and hesitation, we commit finally to somebody or something. Once we've done that we have to shut the oven door and cook. Commitment means we haven't left ourselves an escape hatch. Any marriage, any committed relationship—including commitment to our children, to our parents, to our friends—is about this kind of choice.

When we "shut the door" are we going to be happy? Some of the time. But that's not the point. The point of commitment isn't whether or not it pleases us. Some of the time it will, of course; but don't count on it.

Commitment is not always to another person. We might make a commitment to being alone. For most people such a commitment is good practice, at least once in a while. Perhaps we commit ourselves to being alone for six months, or a year, or five years. Few of us see being alone as just being alone; we see it as loneliness and misery. Yet I'm not talking of some kind of withdrawal into a cave. I mean that in being alone we can practice devoting ourselves to everything and everyone. If we do such a practice we must be honest about the reservations that will accompany it. Nobody wants to devote themselves to everything and everyone. It's a gut-level, demanding practice that not many are eager to do.

Jesus said, "Even as ye have done it unto the least of these, ye have done it unto me." We can't be committed to anything or anyone unless we're committed to everything. That doesn't mean we have to like it, or that we can totally do it. But that is the practice. It's important for each of us to recognize what, in our own life, is "the least of these." We immediately think of people who are very poor. But "the least of these" is what is least to *me*, or to *you*. What is least to you? What in your life are you least interested in serving? For most of us "the least of these" are certain people we dislike or have trouble with: people we consider dispensable. "The least of these" can also be people we're afraid of, people we fear to be around. More subtly, they may be those we feel we must instruct, or enlighten, or help.

You may argue, "Let's be realistic. How can I possibly devote myself to someone I can't even stand? In fact, if I'm within thirty feet of him, it's too much." How can we do that? Well, we learn to practice with it. That means absolute honesty with ourselves: recognizing that we don't like that person and don't want to be around him, and of course observing of all the emotional thoughts surround the relationship. We use this approach with our jobs, also. Some of us are working as tasks that we feel are beneath us (whatever on earth that could mean). "I've got a college degree. Why am I just putting boxes on a shelf? How can I possibly devote myself to such a menial job?"

People want practice to be nice, to be easy. But it's not easy. It's not hard to say "Oh, I'm committed to the world, to the dharma." But it's very hard to *do*. The world, the dharma, is revealed to us by each creature and thing we encounter. Are we committed to that street person vomiting in the gutter? Are we committed to the clerk who just shortchanged us, or to the person who acts superior to us?

Since we are Buddha-nature or truth, we know that joy is our birthright. Where is it? It's waiting for us in the very practice we're talking about. Only through such practice can we move into joy or true commitment in our work, in our relationships, in all of our life.

Because our major difficulties are with people, we don't talk as much as we might about our commitment (or lack of it) to objects. For example: if we keep our room a mess, we're not committed. We are indicating that there's something more important to us than the objects which are our life. (I was brought up by a perfectionist mother, and for a number of years I rebelled against that by being as sloppy as could be.) We're not talking about neurotic neatness, either. Nevertheless, our practice should embrace each person and thing, each cat, each light bulb, each piece of sandpaper, each vegetable, each diaper. If we don't take good care, then we won't know what commitment is. Commitment isn't something that happens by chance. Commitment is a capacity. And it grows as a muscle grows: by being exercised.

I don't want to be laying out a new set of "shoulds." I don't talk much about the Precepts because people misinterpret them: "I should be neat. Joko says so." But we need to look at our tendency to throw things around, to leave lights burning unnecessarily, to take more on our plate than we can eat. Why? If our commitment isn't total, then what we call our commitment to our marriage, to our child, to our work, to our practice, to the dharma, will be undermined. "Even as ye have done it unto the least of these, ye have done it unto me." If we want to know joy, we can't say, "Oh well, I'm just careless." Our practice is always with "the least of these."

Commitment is a functioning. Because we try to avoid function-ing, the witness has to be as sharp as a tack. I don't care how many enlightenment experiences you cling to. There's nothing but daily life. *This table is the dharma.* Yesterday it was dusty; today, it's been dusted. We're coming to the end of sesshin. But don't fool your-self: the hard sesshin begins as you reenter your normal schedule.

IX. SERVICE

Thy Will Be Done

This week many of us watched a television documentary on the life and works of Mother Teresa. Some call her a saint. I doubt that such a title means much to her; but what I found most remarkable was that she was just doing the next thing and the next thing and the next thing, totally absorbing herself in each task—which is what we need to learn. Her life is her work, doing each task wholeheartedly, moment after moment.

We sophisticated Americans have difficulty comprehending such a way of life; it's *very* difficult, and yet it is our practice. Not my will but Thine be done. This does not mean that *Thine* is other than myself, but it *is* other in this sense: my life is a particular form in time and space but *Thine* (Thy Will) is not time or space but their functioning: the growing of a fingernail, the cleansing done by the liver, the explosion of a star—the agony and wonder of the universe. The Master.

A problem with some religious practices is the premature attempt of individuals to practice a life of "Thy Will be done" before there is any comprehension of what is entailed. Before I can understand *Thy Will*, I must begin to see the illusion of *my will*: I must know as thoroughly as possible that my life consists of "I want" and "I want" and "I want." What do I want? Just about anything, sometimes trivial, sometimes "spiritual," and (most usually) for you to be the way I think you should be.

Difficulties in life arise because what *I want* will always clash sooner or later with what *you want*. Pain and suffering inevitably follow. In watching Mother Teresa it is obvious that where no *I want* exists there is joy; the joy of doing what needs to be done with no thought of *I want*.

One point she makes is the difference between one's work and one's vocation. Each of us has some form of work—as a doctor, lawyer, student, homemaker, plumber—but these are not our vocation. Why? The dictionary tells us that "vocation" is from the Latin *vocatio*, to call or summon. Each of us (whether or not we are aware of it) is summoned or called by our True Self (Thine); we wouldn't be at a Zen center if something were not stirring. The life of Mother Teresa is not to serve the poor, but to respond to that summons or call. Serving the poor is not her work, it is her vocation. Teaching is not my work, it is my vocation. And the same for you.

Actually our work and our vocation are one. Marriage, for instance, is many kinds of work (earning income, caring for children and the home, serving the partner and the community), but the vocation of marriage remains the Master. It is our true self, calling, summoning ourselves. When we are clear as to who is the Master, the work flows easily. When we are not clear our work is flawed, our relationships are flawed, any situation in which we participate is flawed.

We all spin merrily onward, doing our work, but we may be blind about our vocation. So how do we grow less blind, how can we recognize our vocation, our Master? How do we understand "Thy Will be done"?

Two practice stages are required (and we vacillate between them). The first is to acknowledge honestly that I don't want to do Thy Will—in fact, forget it, I have no interest in doing it. I want to do what I want nearly all the time: I want to get what I want; I want nothing to be unpleasant for me; I want success, pleasure, health, and nothing else. This sense of *I want* pervades every cell of our bodies; it is impossible for us to conceive of life without it.

Still, as we patiently sit over the years with as much awareness as we can muster, a second stage is building: knowledge builds in our cells of who we truly are and, at the same time, our conceptual beliefs (my will) slowly weaken. Some people like to see Zen practice as an esoteric, removed, separate reality. No, not at all. Instead a slow shift at the cellular level teaches us year after year. Without philosophical pondering, we begin to see who is the *Master. Thy Will* and *my will* are more and more one and the same.

I don't feel sorry for Mother Teresa. She does what gives her the greatest joy. I am sorry for all of us who are blindly stuck in a life of *my will be done*, stuck in anxiety and turmoil.

All of our lives bring problems—or are we given opportunities? Only when we have learned how to practice and can choose not to escape our opportunities but to sit through our anger, resistance, grief and disappointment can we see the other side. And the other side is always: not my will but Thine be done—the life we truly want. What is necessary? A lifetime of practice.

No Exchange

What is the distinction between a life that is manipulative and a life that is nonmanipulative? As Zen students we probably don't think of ourselves as manipulative. Granted, we're not hijacking airplanes. But in a subtle sense we are all manipulative, and we really don't want to be that way.

Let's consider two ways in which the action of our life might take place. On the one hand the action might be dictated by our "false mind": the mind of opinions, fantasies, desires—the little mind we encounter when we sit. For example, we might for some reason dislike a man, and so treat him in a biased manner. On the other hand action might come from the sensory input to our life. Suppose that in crossing the kitchen I drop a grape on the floor. I notice it, bend down, pick it up. That is action dictated by sensory input; it is nonmanipulative action.

But suppose that I have a concept: the kitchen *must be clean*. Having this concept I look for ways to clean up the kitchen. Now, it's OK to have this concept; it's fine to have a clean kitchen. But when the concept is not seen *as* a concept—for instance, if we live in a family where having a clean house dominates family life—we have a concept producing actions instead of actions coming from perceived need. For instance, the degree of cleanliness of the kitchen floor will probably be dictated by whether or not you have young children. If you have three or four children under the age of six, your kitchen floor is not going to be spotless—unless you're the kind of mother who thinks that a shining kitchen is more important than family. And some of us have grown up in families like that. In cases like this something is backwards. A concept is not seen as merely a concept—it's seen as the Truth. "Kitchens should be clean. It's *bad* if kitchens are not clean."

To fulfill our concepts we'll ruin families, nations, anything. All wars are based on concepts, some ideology that a nation says is the Truth. False mind is always dictatorial, always wants to fix the world to enforce the concept instead of being open to perceived need. So when action is backward it becomes manipulative. We must have concepts in order to function; they are not the problem. The problem is created when we believe they are the Truth. Thinking a kitchen *has* to be clean isn't the Truth—it's a concept. False mind deals in *exchange*, not experience. What does that mean?

Our suffering is rooted in a false sense of self, a self composed of concepts. If we think this self truly exists, and believe its concepts are the Truth, then we begin to feel we must protect that self, must fulfill its desires. If we think a kitchen must be clean, then we exert ourselves to carry out this dictate, even bullying others to achieve our end. A "self" is simply a person who believes his or her concepts are the Truth, and is obsessed with doing anything possible to protect the self with concepts to promote its pleasure and comfort.

When we live in this way two words rule our universe: *I want*. If we really look we'll find that *I want* is running our lives. We may want approval, we want success, we may want to be enlightened, we may want to be quiet, we may want to be healthy, we may want excitement, we may want to be loved. "I want, I want, I want, I want." And always we want because we're trying to care for this concept we think is our "self." We want to make life fit our concepts of it.

For instance, if we want to appear as an unselfish person, we'll set everything up so that we'll look selfless. (It probably has nothing to do with being so.) Then no deed, no action, none of our behavior is free of the expectation of an *exchange*. When we do an action we expect something back. In exchange for what we do we expect a return. In ordinary exchange, if you're selling bananas and I give you so much money, I'll get so many bananas; that's true exchange. But the game we play of expecting an exchange for our deeds is not like that.

For instance, if I give a gift of time, money, or effort, what do I expect? What do you expect? Perhaps I feel I am entitled to gratitude. If we give anything we expect an exchange of *something*. We want that person out there to fulfill our personal concepts. When we give a gift we're noble, right? We're *giving* him something; shouldn't he at least notice? We expect an exchange for that. It's a barter; we've turned life "out there" into something with which we're bartering.

If we work for an organization, we expect an exchange for that. If we do something for the organization, where's the other half of it, where is the exchange? If we join an organization, we expect something: perhaps recognition, importance, special treatment.

If we're patient in a difficult situation, and hold our tongue ("You know, anybody else would really explode, but *I'm* patient"), what is the exchange we expect then? *Someone* should notice how patient we've been! Always we're looking for the exchange; we might as well put a dollar sign on it. Or if we're understanding and forgiving ("After all, anyone *knows* how difficult she is") we expect what? If we *sacrifice* ourselves, what is the exchange we should get? Many parent-child games are in this area. "I've done everything for you—you ungrateful so and so!" This is the "exchange" mentality: manipulative, a subtle form of hijacking.

What we expect we rarely get. If we practice long enough we come to see that any expectation of exchange is in error. The world does not consist of objects "out there," whose purpose is to fulfill my concepts. And in time we see more clearly that almost everything we do is with an expectation of exchange—a most painful realization.

When expectation fails—when we *don't* get what we're after—at that point, practice can begin. Trungpa Rinpoche wrote that "Disappointment is the best chariot to use on the path of the Dharma." Disappointment is our true friend, our unfailing guide; but of course nobody likes such a friend.

When we refuse to work with our disappointment, we break the Precepts: rather than experience the disappointment, we resort to anger, greed, gossip, criticism. Yet it's the moment of being that disappointment which is fruitful; and, if we are not willing to do that, at least we should notice that we are not willing. The moment of disappointment in life is an incomparable gift that we receive many times a day if we're alert. This gift is always present in anyone's life, that moment when "It's not the way I want it!"

Since daily life moves quickly, we don't always have clear awareness of what's happening. But when we sit still we can observe and experience our disappointment. Daily sitting is our bread and butter, the basic stuff of dharma. Without it we tend to be confused.

After even a short sesshin such as we had last weekend, it's gratifying to me to see the softening, the opening in people. And sesshin is simply a refusal to meet our expectations! From beginning to end a sesshin is designed to frustrate us! Inevitably, it gives us some pain, mental or physical. It's a prolonged experience of "It's not the way I want it!" When we sit with that, there's always a residue of change left in us. In some cases it's very obvious. But the people who do best in sesshin are usually the ones who have not sat many sesshins. The oldtimers can avoid sesshin even as they do it! They know how to avoid leg pain so that it doesn't get *too* bad; they know many subtle tricks to avoid the whole thing. Because newcomers are less skilled, sesshin hits them harder and there's often an obvious change.

The more we are aware of our expectations, the more we see that our urge is to manipulate life rather than live it just as it is. Students whose practice is maturing aren't angry as often because they see their expectations, their desires, before they produce anger. But if the stage of anger is reached, it *is* practice. Our signal to practice, our "red light," is at the point of upset, the disappointment. "It's not the way I want it!" Some expectation has not been fulfilled and we sense the irritability, frustration, the desire to have it otherwise. "I want" has been frustrated. This very point when "I want" has been frustrated is the "gateless gate" — because the only way to transform "I want" into "I am" is to experience one's disappointment, one's frustration.

Action that comes from experience — picking the grape off the floor — is action coming from perceived need; it is nonmanipulative. Action that comes from the false mind of expectation, "I want," is tyrannical, from a hijacker's mind. When we believe our thoughts and concepts about other people or events we tend to be manipulative, and our life has little compassion. But a life of compassion is nonmanipulative — a life of no exchange.

The Parable of Mushin

Once upon a time, in a town called Hope, there lived a young man called Joe. Joe was much into dharma studies, and so he had a Buddhist name. Joe was called Mushin.

Joe lived a life like anyone else. He went to work and he had a nice wife; but, despite Joe's interest in the dharma, Joe was a macho, know-it-all, bitter guy. In fact he was so much that way that one day, after he'd created all sorts of mayhem at work, his boss said, "I've had enough of you, Joe. You're fired!" And so Joe left. No job. And then when he got home he found a letter from his wife. And she said, "I've had enough, Joe. I'm leaving." So Joe had an apartment and himself and nothing else.

But Joe, Mushin, was not one who gave up easily. He vowed that although he didn't have a job and wife, he was going to have the one thing in life that really mattered — enlightenment. And off he rushed to the nearest bookstore. Joe looked through the latest crop of books on how to achieve enlightenment. And there was one that he found especially interesting. It was called *How to Catch the Train of Enlightenment*. So he bought the book and pored through it with great care. And when he'd studied it thoroughly he went home and gave up his apartment, put all his earthly belongings in his backpack, and went off to the train station on the edge of town. The book said that if you followed all its directions—you do this, and do that, and you do that—then when the train came you'd be able to catch it. And he thought, "Great!"

Joe went down to the train station, which was a deserted place, and he read the book once again, memorizing the directions, and then settled down to wait. He waited and waited and waited. Two, three, four days he waited for the Train of Enlightenment to come, because the book said it was sure to come. And he had great faith in his book. Sure enough, on the fourth day, he heard this great roar in the distance, this enormous roar. And he knew this must be the Train. So he got ready. He was so excited because the Train was coming, he could hardly believe it . . . and . . . *whoosh* . . . it went by! It was only a blur, it went by so fast. What had happened? He couldn't catch it at all!

Joe was bewildered but not discouraged. He got out his book again and studied some more exercises, and he worked and worked and worked as he sat on the platform, putting everything he had into it. In another three or four days he once again heard a tremendous roar in the distance, and this time he was certain he would catch the Train. All of a sudden there it was . . . *whoosh* . . . it was gone. Well, what to do? Because obviously there was a train, it wasn't as though there was no train. He knew that, but he could not catch it. So he studied some more and he tried some more, he worked and worked, and the same thing happened over and over again.

As time went on other people also went to the bookstore and bought the book. So Joe began to have company. First there were four or five people watching for the Train, and then there were thirty or forty people watching for the Train. The excitement was tremendous! Here was the Answer, obviously coming. They could all hear the roar as the Train went by and, although nobody ever caught it, there was great faith that somehow, some day, at least one of them would catch it. If even one person could catch it, it would inspire the rest. So the little crowd grew, and the excitement was wonderful.

As time went on, however, Mushin noticed that some of these people brought their little kids. And they were so absorbed in looking for the Train that, when the kids tried to get mom and dad's attention, they were told "Don't bother us, just go play." These little kids were really being neglected. Mushin, who was not such a bad guy after all, began to wonder, "Well, gee, I'd like to watch for the Train, but somebody's got to take care of the kids." So he began to devote some time to them. He looked in his back-pack and took out his nuts and raisins and chocolate bars and passed all this stuff out to the kids. Some of them were really hungry. The parents who were watching for the Train didn't seem to get hungry; but their kids were hungry. And they had skinned knees, so he found a few bandaids in his backpack and took care of their knees, and he read them stories from their little books.

And it began to be that while he still took some time for the Train, the kids were beginning to be his chief concern. There were more and more of them. In a few months there were also teenagers, and with teenagers there is a lot of wild energy. So Mushin organized the teenagers and set up a baseball team in back of the station. He started a garden to keep them occupied. And he even encouraged some of the steadier kids to help him. And before you knew it he had a large enterprise going. He had less and less time for the Train and he was angry about it. The important stuff was happening with the adults waiting for the Train, but *he* had to take care of all this business with the kids, and so his anger and his bitterness were boiling. But no matter what, he knew he had to care for the kids, so he did.

Over time, hundreds and thousands of Train watchers arrived, with all their kids and relatives. Mushin was so harried with all the needs of the people that he had to add on to the train station. He had to make more sleeping quarters; he had to build a post office and schools and he was *busy*; but his anger and his resentment were also right there. "You know, I'm only interested in enlightenment. Those other people get to watch the Train and what am I doing really?" But he kept doing it.

And then one day he remembered that while he'd thrown out most of the books in his apartment, for some reason he had kept one small volume. So he fished it out of his backpack. The book was *How to Do Zazen*. So Joe had a new set of instructions to study. But these didn't seem so bad. He settled down and learned how to do zazen. Early in the morning before everyone else was up, he'd sit on a cushion and do this practice for a while. And over time this hectic, demanding schedule in which he had unwillingly become immersed didn't seem so much of a strain for him. He began to think that maybe there was some connection between this zazen, this sitting, and the peace he was beginning to feel. A few others at the station were also getting a bit discouraged about the Train they couldn't catch; so they began to sit with him. The group did zazen every morning and, at the same time, the Train-watching enterprise kept expanding. At the next train station down the tracks there was a whole new colony of train watchers. The same old problems were developing there, so sometimes his group would go there and help in straightening out their difficulties. And there was even to be a third train station . . . endless work.

They were really, really busy. From morning till night they were feeding the kids, doing carpentry, running the post office, setting up the new little clinic—all that a community needs to function and survive. And all this time they weren't getting to watch for the Train. It just kept going by. They could hear the roar. And some jealousy and bitterness were still there. But still, they had to admit, it wasn't the same anymore; it was there and it wasn't there. The turning point for Mushin was when he tried something described in his little book as "sesshin." He got together with his group and, in the corner of the train station, they set up a separate space and for four or five days they would steadily do zazen. Occasionally they'd hear the roar of the Train in the distance, but they ignored it and went on sitting. And they also introduced this hard practice to the other train stations.

Mushin was now in his fifties. He was showing the effect of the years of strain and toil. He was getting bent and weary. But by now he no longer worried about the things he used to worry about. He had forgotten the big philosophical questions that used to grip him: "Do I exist?" "Is life real?" "Is life a dream?" He was so busy sitting and working that everything faded out except for what needed to be done every day. The bitterness faded. The big questions faded. Finally there was nothing left for Mushin except what had to be done. But he no longer felt it had to be done, he just did it.

By now there was an enormous community of people at the train stations, working, bringing up their children, as well as those who were waiting for the Train. Some of those slowly were absorbed back into the community and others would come. Mushin finally came to love the people watching for the Train, too. He served them, helped them to watch. So it went for many years. Mushin got older and older, more and more tired. And his questions were down to zero. There were none any more. There was just Mushin and his life, doing each second what needed to be done.

One night, for some reason, Mushin thought, "I will sit all night. I don't know why I want to do it. I'll just do it." For him sitting was no longer a question of looking for something, trying to improve, trying to be holy. All those ideas had faded years ago. For Mushin there was nothing except just sitting: Hearing a few distant cars at night. Feeling the cool night air. Enjoying the changes in his body. Mushin sat and sat through the night, and at daybreak he heard the roar of the Train. Then, very gently, the Train came to a stop exactly in front of him. He realized that from the very beginning he had been on the Train. In fact he was the Train itself. There was no need to catch the Train. Nothing to realize. Nowhere to go. Just the wholeness of life itself. All the ancient questions that were no questions answered themselves. And at last the Train evaporated, and there was just an old man sitting the night away.

Mushin stretched and arose from his cushion. He went and fixed morning coffee to share with those arriving for work. And the last we see of him, he's in the carpentry shop with some of the older boys, building a swing set for the playground. That's the story of Mushin. What was it Mushin found? I'll leave that to you.

Notes

PREFACE

v. **an amazingly pure** Alan Watts, "Beat Zen, Square Zen and Zen" in *This Is It and Other Essays on Zen and Spiritual Experience* (New York: vintage, 1973), 106.

vii. **There is another** Elihu Genmyo Smith, "Practice: Working With Everything," unpublished manuscript.

viii. **From the withered** Compare *The Record of Tung-Shan*, translated by William F. Powell (Honolulu, Hawaii: University of Hawaii Press, 1986), 63: "The blooming of a flower on a sear old tree, a spring outside of kalpas." See also Isshu Miura and Ruth Fuller Sasaki, *Zen Dust: The History of the Koan and Koan Study in Rinzai (Lin-Chi) Zen* (New York: Harcourt, Brace & World, 1966), 104: "In the spring beyond time, the withered tree flowers."

THE FIRE OF ATTENTION

32. **If you can** Huang Po, from *The Zen Teaching of Huang Po*, translated by John Blofeld (New York: Grove Press, 1959), 33.

33. **We cannot solve** Compare *The Recorded Sayings of Ch'an Master Lin-chi Hui-chao of Chen Prefecture*, translated by Ruth Fuller Sasaki (Kyoto, Japan: The Institute for Zen Studies, 1975), 9 ff.

33. **On no account** Huang Po, in Blofeld, *The Zen Teaching*, 130.

NO HOPE

67. **Next, you should** Zen Master Dōgen and Kōshō Uchiyama, *Refining Your Life: From the Zen Kitchen to Enlightenment*, translated by Thomas Wright (New York, Tokyo: Weatherhill, 1983), 5.

67. **He carried a** *Ibid.*, 9 ff.

LOVE

71. **When, through practice . . . own conditioned preconceptions.** Menzan Zenji, *Shikantaza: An Introduction to Zazen*, edited and translated by Shōhaku Okumura, published by Kyoto Sōtō-Zen Center (Toyko, Japan: Tōkō Insatsu KK, 1985), 106.

TRUE SUFFERING AND FALSE SUFFERING

107. **This mind is** Huang Po, in Blofeld, *The Zen Teaching*, 33.

RENUNCIATION

110. **Renunciation is not** Shunryu Suzuki, Roshi, *Wind Bell 7*, no. 28, 1968.

TRAGEDY

119. **a man was** "A parable," in *Zen Flesh, Zen Bones: A Collection of Zen and Pre-Zen Writings*, compiled by Paul Reps (Garden City, New York: Anchor Books, no date), 22 ff. Compare also Leo Tolstoy, "My Confession," in *The Complete Works of Count Tolstoy*, vol. 13, translated and edited by Leo Weiner (Boston: Dana Estes & Co., Publishers; Colonial Press, 1904), 21 ff.

THE OBSERVING SELF

122. **Who is There?** in Arthur J. Deikman, M.D., *The Observing Self: Mysticism and Psychotherapy* (Boston: Beacon Press, 1982), 88. See also 91–118, *passim*.

GREAT EXPECTATIONS

149. **Let go of** Compare "Shōji," final paragraphs, *A Complete English Translation of Dōgen Zenji's Shōbōgenzō*, translated by Kōsen Nishiyama and John Stevens (Toyko, Japan: Kawata Press, 1975), 22.

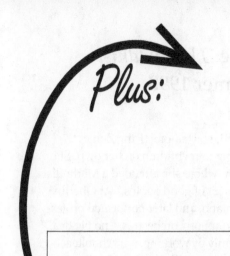

Plus:

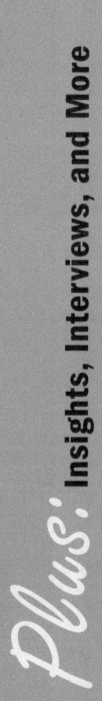

Plus: Insights, Interviews, and More

From Tricycle: *The Buddhist Review*, Summer 1998

Charlotte Joko Beck, 81, started practicing Zen in the mid-sixties after raising four children on her own. She grew up in New Jersey, where she attended a Methodist church and "learned a lot of good quotes." At Oberlin College, she studied piano, and later performed professionally "with little symphony orchestras—no big deal." She supported her family by working as a schoolteacher, a secretary, and finally as an administrator in the chemistry department at the University of California, San Diego. When she retired in 1977, she went to live at the Zen Center of Los Angeles.

In 1983, the Zen Center of San Diego opened—in "two little houses, side by side, no sign"—with Joko as teacher. She's evolved her own way of teaching, which is always open to change.

"I'll pick up anything if it's useful. It's a question of seeing what really transforms human life. That's what we're interested in, isn't it?" She's just discovered Pilates, a form of exercise combining yoga, dance, and resistance training, and "probably I'm learning something there that will get mixed in, too."

Why did you start practicing?

I had a fine life. I was divorced—my husband was mentally ill—but I had a nice man in my life. My kids were okay. I had a good job. And I used to wake up and say, "Is this all there is?"

Then I met Maezumi Roshi, who was a monk at the time. He was giving a talk in the Unitarian Church downtown. I was out for the evening with a friend, a woman, a sort of hard-boiled business type, and we decided to hear his talk. And as we went in, he bowed to each person and looked right at us. It was absolutely direct contact. When we sat down, my friend said to me, "What was *that?*" He wasn't doing anything special—except, for once, somebody was paying attention.

I wanted whatever he had. I found a sitting group of two in San Diego, and I became the third. Maezumi would come down once in a while. Eventually, I began to go up to Los Angeles every week or two for the sittings. And occasionally I began doing sesshins with Soen Roshi.

At one point I had a little breakthrough or "opening"—which I now think is a waste oftime, but at the time, I thought it was important.

What was your "little breakthrough"?

Just a sense of everything being whole and complete, with no time or space—no "me"—that sort of thing. It was terrifying! And I was furious. I went in and threw something at Soen Roshi. [laughs]. He ducked. I said, "You mean to say we sit and struggle and struggle just to realize this—that there's really just nothing at all?" Because I didn't really understand. He said, "Well, it's not terrible; it's just astonishing."

You think it's a waste of time to have a breakthrough?

Not a waste of time, but it's not the point. It doesn't mean you know what to do with your life. You can sit for twenty years and be wasting your time.

What I'm interested in is the process of awakening, the long process of development, which may, or may not, have breakthroughs as natural fruit. What genuinely concerns me is the necessity for a student to learn to be as awake as possible in each moment. Otherwise, it can seem as if the point of practice is to have breakthroughs.

I've spent years thinking about this, and seeing how it's ordinarily done, and I'm just saying there's a way to teach so that people learn to use their daily life as practice—as the key to awakening. And that's how we do it here.

How did you start this center? (The Zen Center of San Diego)

A group of people got together and bought the two houses, next door and this one. I had to have a place to live, and you need a place to sit, and I really wanted a little separation. We juggled space, and it isn't ideal, but that's part of why this place is interesting. Nothing has ever been quite right, but we learn to make do and make that our practice. A little chaos is often useful.

How do people practice?

Basically, new students usually learn to experience their body and label their thoughts. I don't mean to analyze thoughts or pick them apart. It's a little like vipassana, but instead of saying, "Thinking, thinking," I like people to just recite their thoughts back. If you do that for three or four years, you'll know a lot about how your mind operates.

How might you label thoughts?

Mmm, "Having a thought about Mary. . . . Having a thought that I really don't like Mary . . . Having a thought that I can't stand the way she bosses everyone around." That's the way we think, right? And in time, as we watch our thoughts our thinking becomes more objective. But most people, instead of just having a thought about Mary, go further: "Gosh, I can't stand her; she really makes me mad." Now they've got an emotion. What we need to learn to do is to see the thought as a thought, and then feel the body tighten. The body is going to tighten if you're angry with somebody, right? So just be the tightening. Forget the thinking at this point, and just be the anger, the tension or vibration. When you do that, you're not trying to change your anger. You're just being with it, totally. Then it is able to transform itself. That's transformation as opposed to change—a critical difference. Religion always is trying to change you: you know, "You're not a good girl; be a good girl." But here, in labeling and experiencing, you're learning to be less emotional, less caught by every passing thing that goes on in your head.

The anger gets a little weaker, a little less demanding, and at some point, you begin to notice the difference. Something that would have made you jump with anger—you can watch it. The observer is beginning to grow. And in experiencing the bodily tension, you're not suppressing the emotion; you're feeling it. You're transforming the dualism of self-centered thoughts, opinions, and emotions into the non-dualism of direct experiencing. So when people come in to talk to me, after a few months I'll probably say, "I want you to bring in an episode that bothers you and tell me how you see that as practice. Suppose somebody yells at you in an unfair way. What is practice?" We work through it, and the next week we do it again with another episode.

Their own personal koan.

Yeah. "I went to a party and my husband spent all his time looking at other women." How do you practice with that?

At first, students may have some ideas that are crazy and lead to even more upset. In time, they learn the difference between getting angry, arguing or shouting—and just experiencing the anger. This doesn't mean we don't take action in a situation—often we do. But we don't do it in anger; we just face the facts of the situation. And we learn to deal with our life without our ego-centered emotions running the show.

How did your way of teaching evolve?

One interest I always had was psychology, and at some point, I had probably read an enormous amount and, in sitting, began to see what to do with it.

Who were you reading?

Karen Horney, for one, but all the standard texts, and slowly I began to evolve a practice that is classical but also therapeutic—though not therapy. I notice that a lot of people think, "Yes, you have a strong Zen practice, but since that won't take care of any emotional problems, you need therapy, too." I have great respect for therapy, but for most people—I'm not talking about disturbed people—I feel that practice can be a complete path.

This is certainly a departure from the way you were trained.

Many Zen practices are about suppression—sheer concentration and shutting out things. I realized that what you shut out is exactly what turns around and runs you. So I began trying to get students to work in a different way, and it proved to be effective. Dogen [the thirteenth-century Zen master] said that to study Buddhism is to study the self. To study the self—your thoughts, etc.—is to forget the self. And to forget the self is . . . what? It's to be enlightened by all things. Suppose somebody has hurt my feelings—or so I think. What I want to do is to go over and over and over that drama so I can blame them and get to be right. To turn away from such thinking and just experience the painful body is to forget the self. If you really experience something without thoughts, there is no self—there's just a vibration of energy. When you practice like that ten thousand times, you will be more selfless. It doesn't mean that you're a ghost. It means that you're much more non-reactive, in the world but not of it. Dualism transforms into non-dualism, a life of direct and compassionate functioning.

You use several images for experiencing pain rather than running from it: stretching out on an "icy couch," moving onto the "razor's edge"—excruciating images.

But they're not excruciating—the minute you experience what you've been running from. For instance, suppose you've been humiliated. Well, nobody likes to be humiliated; it's one of the yuckiest feelings in the world. We want to pretend it didn't happen; we want to blame someone. To turn around and just to feel that—eech. But part of what sitting does, in time, is give you the strength to stay with it. And after a while—surprise!—it's okay. And then it's not only okay, but it begins to change things. It's as if the sun comes up.

See, that *is* the gate to enlightenment. When you practice like that thousands of times, you're a different person. There's a true transformation, and that's what practice is about.

Your experience is that whatever the emotion or experience, the fact of *being* in it one hundred percent turns into joy?

Right. Because it wipes out self. There's nothing left but openness. Not happiness, but openness—joy. Joy can also be sad. Perhaps you've had a grandmother die. She dies peacefully, and in a way it's wonderful. It's time—there's no conflict about it, and there's even joy, because that's the way it is, and it's fine.

In your talks, you refer to Christianity, Sufism, psychology—almost as often as you cite Zen sources. That seems like a step in the direction of

Americanizing Zen—America as the idea of assimilating varied sources.

It's fine for Zen to be Americanized. But that doesn't just mean it must *look* American. It means to practice in a way that's best in this culture. Here, even though we can get badly stuck in our intellectual approaches, we need a base of that sort—but it needs to be handled carefully, or the tail will wag the dog. Anyway, I have nothing against centers where everyone who sits wear a robe. I have nothing against Japanese-type services—they're very beautiful. It's just not where my energy goes. What I'm interested in is, how do we all learn? How do we transform?

You describe yourself as a very persistent student. What was driving you?

I have always been determined. I am determined about Pilates, and I am determined about anything I do. I'm not saying it's a virtue. I'm just saying that's how it is. But everyone has something that pulls them.

For some people it could be the magic belief in transformation, or the expectation of happiness—all those things you're constantly warning your students against.

Everyone starts practice with many false beliefs. We all start by thinking that something is going to fix us and make us feel better and more protected.

How does someone begin to see emptiness?

I don't think you see it. You have to *be* it. Emptiness simply means an absence of reactivity.

When you relate to somebody, there's not you and me and your little mind running its little comparisons and judgments. When those are gone, that is emptiness. And you can't put it into words. That's the problem for people.

They think there's some way to push for an experience such as emptiness. But practice is not a push toward something else. It's the transformation of your self. I tell people, "You just can't go looking for these things. You have to let this transformation grow." And that entails hard, persistent, daily work. I simply wouldn't let an irritable thought go through my mind without noting, "Oh, that's interesting. What's going on here?" I don't mean analyzing it, but just stopping. There has to be that ability to stand back and say, "Yeah, interesting that I do that." Right there. I may go back to it if I'm busy talking to you. But it's been registered. I'm not going to let that one go by; it's too interesting. It's not good or bad. It's just interesting to note that you do that.

You keep talking about strengthening the observer, the witness.

The truth is, we don't want to observe, particularly if we're upset. What do we want to do? We want to be upset, because then I remain the center of the drama. If I observe, I weaken that self-centered position.

Does the witness ever fade?

The witness that observes fear, say, finally has to go into just *being* fear, just feeling fear and abandoning as much as possible the thoughts connected with it. Easy to say, hard to do. It's the same thing for a musician. There's a lot of self-conscious practice. But there are times when there's *just* playing and you don't even know you're playing. There's no witness, there's nothing: just functioning. That's the stage of the witness fading, just by growing selflessness. You're no longer living your life to reach a goal, but just for the sake of living it. And that can include working very hard on a project, but there's no ambition. No goals, and yet goals are steadily accomplished. There's just living, enjoyment of life.

Why do you think so many people are turning to Zen, to Buddhism, now?

Because people think something or someone will relieve them of their pain and disappointment. "I'll try Zen." But it's only when you understand why this attempt is backwards that you can seriously begin a real practice. Not to get rid of pain and disappointment, but to put yourself right into them.

You're very enthusiastic about the uses of disappointment.

See, we usually live our lives out of the ceaseless hopes and expectations of this self-centered mind or ego. And if that works, if you're unfortunate enough that it works—you want the ideal man, you get the ideal man; you get the ideal job; everybody loves you—then you forge ahead in your usual way until something comes along that stops you in your tracks. Usually, it's a disappointment or disaster of some sort. What most people do then, naturally, is try harder. They want to be happy, so they look for a new formula, and that's when they take up some sort of a practice, or go to church, or do something.

If you're lucky, though, you continue to meet painful disappointment. "Gosh, it just doesn't work; I don't know what to do next—I'm baffled." I always congratulate people who arrive at this crossroads—"Aren't you lucky!"—because now the true path can be glimpsed. A real practice can begin. It doesn't mean that if I get disappointed, I like it. But I know it now for what it is.

In the first years of practice, you say, there's often a movement from unhappiness to happiness. What's happening?

The early years increase objectivity. The dominance of self-centered emotion (particularly "poor me") is challenged; the body is more stable and strong. And at some point, under the pressure of practice and life's inevitable disappointments, a turning point is reached. Resistance weakens, and we are more willing to investigate the ceaseless desires of the ego. You know the classic Zen vow:

"Desires are inexhaustible; I vow to put an end to them." Well, you can't *want* to put an end to desire—that would be just another ego project—but we can persistently practice: not with ambition "to get somewhere," but with the one true desire that our practice benefit not just ourselves but all sentient beings.

You keep insisting on the importance of daily practice.

It isn't something that we want to do. But that's what keeps you getting more *porous* all the time. A lot of people, though, even though their practice is going well, will read a book, and come in and say, "You know, Joko, in the three or four years that I've worked with you, everything in my life has transformed. I have a happy marriage now, I'm getting along at work, everything is totally different, and I feel much freer. But what about the *real* thing?" They mean, "When do I get enlightened?" And of course, they are walking the path of enlightenment—non-reactivity—but their understanding is partial. In time, if the work continues, they'll see that the process itself *is* the real thing; it *is* the gateless gate.

The capacity for great satori—that's not going to happen to the average person. The person who has a great opening is somebody who has been almost selfless for years, and maybe that last little chunk drops off. I mean, great satori doesn't just fall out of the skies.

Of course, books like *The Three Pillars of Zen* can lead you to hope that enlightenment comes to everyone eventually—and with a big bang.

Plus: **Insights, Interviews, and More**

I know, but mostly, these breakthroughs just increase the ego. A student may come to me and say, "Joko, this thing happened to me. I was taking out the garbage, and you know what?" I say, "Well, I'll give you one minute to tell me" [laughs]. So they tell me—they have their minute. Then I add, "Okay. And how are you and your wife getting along?" Their "little moment" just means they're beginning to be able to be present, if only for a few seconds. But that doesn't necessarily mean they now know how to be with the criticism of a wife or anyone else.

So this process functions kind of as marriage counseling, family therapy?

Not at all—it's training in the process of practice. I'm not telling them, "You should be this or do that." What I may say is, "If you don't buckle down to a serious practice, instead of just talking about it, your marriage will probably end—with pain for many people." And fortunately, they usually do that—and hopeless marriages become good marriages. At least, we haven't lost a marriage yet. Instead of wanting the marriage to meet their own expectations, they begin to really experience their own anger and neediness, and the transformation to harmony begins. It's hard practice! And, of course, wonderful. There are therapeutic outcomes, but it's always practice. Always practice.

In your books, you keep saying that "the problem is never other people."

Never.

What about a woman whose husband swats at the kids, at her?

Out. Leave. Physical abuse, you get out. You don't need to say, "I'll never see you again." But with any physical abuse, just get out.

But wouldn't you say he's the problem here?

Instead of answering directly, let me tell you what I have all students do within the first three to six months. I have them make three lists. And it's fun; you could do it yourself. The first list is: "As a small child, what I was trained to be was. . . ." For instance, I was trained to be "perfect." Never could show anger. Had to succeed at everything, get straight A's, please everybody—the quintessential good little girl. We all sort of know what we were trained to be.

The second list is: "Right now, as an adult, what I require myself to be is. . . ." This may look as if it's your list, but it really isn't, since we use much of our first list—how we were trained—to form our ideas of how we should be now. And the list in itself may be fine—I require myself to be thoughtful, kind, patient, selfless, non-angry: the usual stuff—but until it's really *your* list, you will have a hidden third list.

Now, the third list is more interesting: It's the negative emotions hidden behind the second list. Suppose I have a good friend sick in the hospital. It's Saturday afternoon and I'm worn out, but still I think I should go see her—because my second-list requirement is what? I should be patient, loyal. . . . And I will go see her. But beneath the appropriate action will be what?

Resentment.

Resentment. My third list. In other words, I'll go to see her partly because I love her, but partly because that's what "good" people do—my first two lists. It's obvious that the transforming practice—so I do what I do for the sake of doing it, not because I should—lies in the third list. It's to experience the bodily tension of resentment without my thoughts of how I should be. That begins to weaken this whole conditioned shebang that we live out of. So, in answer to your question, the other person is never the problem. Again, suppose somebody mistreats you. Suppose someone tells lies about you at work. You could say, "I'm upset because she's undermining me—I could lose my job." But what you're really upset about is that your requirement that life be fair—I should be fair, you should be fair—is being attacked. The problem isn't what she does. The problem is that she's attacking your second-list requirement, removing your cover so that you have to feel the unpleasant emotions of the third list. She's brought your fear and anger into the open—exactly what all of us dislike.

And if these lies are truly threatening your job or reputation?

We're not talking about Miss Milquetoast here. We're saying that if you really see your thoughts and experience your anger, you can then act without as much anger, maybe none. You could have lunch with her and say, "You know, we've had a good relationship and I do value you, and yet it's come to my attention. . . . I wonder if we could talk about that. I wonder how you see that." If you can speak without anger, it will be a very different ball game. Do you see what I mean?

You're saying life isn't a problem.

Exactly, it's you who's a problem. It's your reactivity. See, if you could really cease being angry with her, you would be a different person, not just with her but in hundreds of situations in which an attack seems to be coming your way. Your life would be more calm, you'd be better for yourself and other people. See, that person isn't a monster. She's a human being who is ignorant, or else she wouldn't be doing what she's doing. And if, as practitioners, our aim is to save all sentient beings—to use a goody-goody-sounding phrase—we want to benefit her in our interaction with her.

This is the enlightenment process. One idea that really hampers us is to believe that people get "enlightened," and then they're that way forever and ever. We may have our moments, and if we get sick and have lots of things happening, we may fall back. But a person who practices consistently over years and years is more that way, more of the time, all the time. And that's enough. There is no such thing as getting it. Who, after all, would be getting it? There are just stages of selflessness.

When you sit now, what does your mind do?

Nothing much. Thoughts come and go, and that's fine—the mind is meant to think. I stay awake as much as I can. That's all. It's no big thing, except perhaps that I'm not all torn up by worrying, "I've got to get better." My mind is fairly quiet, so I'm not bugged by my mind. I am the way I am, and that's okay.

Did you start out being very reactive?

Oh, I used to throw things at my husband. Bricks. Bricks at windows. Good girls have a lot of anger, you know. And when something sets it off, BOOM, it explodes. I wouldn't say I was chronically like that, but sure, the anger was there all the time.

How does practice create transformation?

You keep experiencing your fear instead of running around it or rationalizing it. You're just afraid. You stay with it and begin to do things that frighten you, not to be virtuous, but so you can directly feel the body sensation we call fear. It's very useful to do things you don't want to do—make that phone call you don't want to make or whatever it is for you. With unfailing kindness, your life always presents what you need to learn.

Whether you stay home or work in an office or whatever, the next teacher is going to pop right up. Let's say you notice that you have no patience with a certain person. Well, right there, you pay attention: "What's this impatience?" As long as you're capable of being annoyed, you can be sure that something will annoy you. When you no longer can be annoyed by little monkeyshines, you'll find most everything agreeable. And of course, you have to watch your own monkeyshines. It's great fun, really. It is! It's fascinating to begin to watch our life unroll and to see what's really going on.

I want to go back to your "little breakthrough." You said, "Breakthroughs aren't the point." But can you get the point without the breakthrough?

What primarily concerns me is the necessity for a student to learn to be as awake as possible in each moment. Otherwise it can seem as if the point of practice is to have breakthroughs. The usefulness of these openings exists only if they clarify life and our ability to live it and serve it. But until mind and body —usually through years of patient practice— cease to *want* an ego-centered life, the openings and their teachings cannot be distorted into ego successes. Only when mind and body are mostly free of reactivity can a true understanding of what life is become possible—not through a momentary breakthrough, but through an open and compassionate living of life.